Large Language Model Recipes

A Hands-On Guide to Fine-Tuning, Optimization, Deployment, and Real-World Applications

Bharath Kumar Bolla
Kalpa Subbaiah
Sashi Kiran Kaata

Apress®

Large Language Model Recipes: A Hands-On Guide to Fine-Tuning, Optimization, Deployment, and Real-World Applications

Bharath Kumar Bolla
Hyderabad, Telangana, India

Kalpa Subbaiah
Hyderabad, Telangana, India

Sashi Kiran Kaata
Glendale, AZ, USA

ISBN-13 (pbk): 979-8-8688-2606-1
https://doi.org/10.1007/979-8-8688-2607-8

ISBN-13 (electronic): 979-8-8688-2607-8

Managing Director, Apress Media LLC: Welmoed Spahr
Acquisitions Editor: Celestin Suresh John
Coordinating Editor: Gryffin Winkler

Cover image by Pixabay.com

Distributed to the book trade worldwide by Springer Science+Business Media New York, 1 New York Plaza, New York, NY 10004. Phone 1-800-SPRINGER, fax (201) 348-4505, e-mail orders-ny@springer-sbm.com, or visit www.springeronline.com. Apress Media, LLC is a Delaware LLC and the sole member (owner) is Springer Science + Business Media Finance Inc (SSBM Finance Inc). SSBM Finance Inc is a **Delaware** corporation.

For information on translations, please e-mail booktranslations@springernature.com; for reprint, paperback, or audio rights, please e-mail bookpermissions@springernature.com.

Apress titles may be purchased in bulk for academic, corporate, or promotional use. eBook versions and licenses are also available for most titles. For more information, reference our Print and eBook Bulk Sales web page at http://www.apress.com/bulk-sales.

Any source code or other supplementary material referenced by the author in this book is available to readers on GitHub (https://github.com/Apress/Large-Language-Models-Recipes). For more detailed information, please visit https://www.apress.com/gp/services/source-code.

If disposing of this product, please recycle the paper

This book is dedicated to my wife, Soujanya, who shouldered the weight of our daily lives so I could focus on these pages, and to our children, Harshith and Daivik, for their endless patience and for allowing their time to be "stolen" for this dream.

—Bharath Kumar Bolla

This book is dedicated to my family, especially my children, Prisha and Vihaan, whose love, patience, and encouragement supported me throughout this journey.

—Kalpa Subbaiah

I would like to express my deepest gratitude to my wife, Ravali, for her constant encouragement and for holding everything together while I focused on writing this book. To my daughters, Bhaavya and Tanvika, thank you for your love, joy, and patience—this work is as much yours as it is mine.

—Sashi Kiran Kaata

Table of Contents

About the Authors

Bharath Kumar Bolla is a highly accomplished Data Science leader with over 15 years of experience, specializing in AI, NLP, and Deep Learning for the past decade. He holds an M.S. in Data Science (the University of Arizona) and an Executive MBA (Product Management).

As an Associate Director at Novartis, he currently drives strategic MLOps initiatives, successfully designing and scaling automated pipelines and deploying cutting-edge Generative AI solutions across multiple European markets. His commercial impact is notable, including architecting a Salesforce recommendation system (5-10% conversion boost) and developing an ML pricing optimization product (+$1.2M revenue at Verizon).

Beyond corporate leadership, Bharath is a prolific academic with over 30 peer-reviewed publications and multiple best paper awards. He is recognized as a "40 Under 40 Data Scientist" (2022) and an "AI Changemaker" and actively supports the community by reviewing AI books and mentoring students. You can know more about him on `https://www.linkedin.com/in/bharathbolla/`.

Kalpa Subbaiah is a leading Data Scientist and AI expert with over 17 years of experience, including more than a decade in Data Science and Machine Learning. She holds a master's degree in Machine Learning and Artificial Intelligence from Liverpool John Moores University and specializes in building end-to-end AI solutions across Azure, Databricks, and AWS.

Kalpa is a GenAI expert, building production-grade LLM and RAG applications, fine-tuning models via Hugging Face, and architecting scalable multi-agent AI systems with robust evaluation frameworks. Her strong experience in finance and manufacturing drives projects in financial AI platforms, smart city solutions, and enterprise analytics, leveraging capabilities such as document intelligence and object detection.

As Vice President and Lead Data Scientist at JPMorgan Chase & Co., she drives large-scale AI/GenAI transformation across the enterprise. Highly certified (Azure Data Scientist/AI Engineer and AWS ML Specialist), she actively delivers global corporate and academic training as a technical trainer and mentor.

Sashi Kiran Kaata is a seasoned cloud data engineer, researcher, and technology leader with over 10 years of experience architecting large-scale data platforms and real-time analytics solutions. Holding a master's degree in Information Science and Technology, he has core expertise in AWS, Snowflake, Databricks, and modern streaming frameworks. At First Citizens Bank, he led enterprise data modernization, designing resilient ingestion and governance architectures, including Data Movement Controls (DMC), which significantly improved platform performance, reliability, and regulatory compliance.

Sashi blends practical engineering with research-driven innovation, actively contributing to the community as a technical conference speaker on distributed systems and scalable cloud architectures. His work extends into blockchain-based workflow modernization, sustainable AI pipelines, and adaptive ETL systems designed to support the next generation of intelligent data platforms. He is a prolific author of numerous peer-reviewed publications on critical areas, including cloud cost optimization, self-healing systems, Green AI, and MicroLLMs.

About the Technical Reviewer

Sanathraj Narayan is a Senior Data Scientist with over a decade of experience in building AI and machine learning solutions, as well as scalable data systems on AWS and Azure. He has held roles at Ericsson, Mindtree, KPMG India, and Cognizant, where he led data-driven projects across the retail, telecom, and consulting sectors. Sanathraj's expertise spans predictive modeling, recommender systems, and the deployment of end-to-end machine learning pipelines. He is also a regular conference speaker, presenting on AI, machine learning, and Generative AI topics.

Acknowledgments

A book like this does not come together without the patience and support of the people closest to you. Before anyone else, we want to thank our families. They gave up evenings, weekends, and more than a few holidays so we could write, revise, and rewrite again. This book would not exist without their encouragement, and we do not take that for granted.

We are deeply grateful to the team at Apress for believing in this project and guiding it from a rough idea to the book you are holding. **Celestin John**, our Acquisitions Editor for AI and Machine Learning, helped shape the direction of this book from the very beginning. His feedback during the early stages pushed us to sharpen our focus and think more carefully about what practitioners actually need. **Nirmal Selvaraj** was our steady point of contact throughout the development cycle, keeping everything on track with dedication and patience. **Gryffin Winkler**, our Coordinating Editor, brought a sharp editorial eye and practical wisdom that made every chapter better. Working with this team was a genuine pleasure.

We also want to extend a special thanks to our technical reviewer, **Sanathraj Narayan**, who read every chapter with the kind of care most authors only hope for. His detailed feedback, grounded in real-world experience across LLMs and applications, caught issues we had missed and raised questions that made our explanations sharper and more honest. The quality of this book improved meaningfully because of his work.

To everyone who contributed to this project, whether named here or not, thank you. Your support, feedback, and commitment helped turn an ambitious idea into something real. We hope the result is worthy of the effort so many people put into it.

Introduction

There is no shortage of LLM books today. Some go deep into the mathematics, leaving you wondering how any of it translates into a working system. Others hand you code snippets and expect you to figure out the rest. We wrote *LLM Recipes* because we wanted a book that does both: one that explains ideas clearly enough to build real intuition and backs every concept up with code you can actually run.

What sets this book apart starts with how it teaches. Rather than jumping straight to code, every major concept is introduced through carefully designed figures, architecture diagrams, and visual walkthroughs. We believe a good diagram can make something click in seconds that paragraphs of text alone struggle to convey. When we explain attention mechanisms, LoRA adapters, or a RAG pipeline, the figure comes first. The code follows once you already have a mental picture of what is happening. Figures like the LLM Data Preparation Workflow (Figure 4-1), the LoRA Weight Decomposition (Figure 8-2), and the RAG Pipeline Architecture (Figure 13-3) are not decorations; they are reference anchors you will return to throughout the book and in your own work.

The second distinction is scope. Most practical LLM guides cover a slice of the problem: a fine-tuning book here, a deployment manual there, a chapter on quantization somewhere else. *LLM Recipes* covers the entire post-training journey in one volume, from prompt engineering and instruction tuning through parameter-efficient fine-tuning, quantization, production deployment, evaluation, RAG, vision language models, and agentic workflows. We wrote it this way because these stages are deeply linked in practice. Your quantization choices affect serving latency. Your instruction format shapes your prompting surface. Your evaluation discipline determines every decision upstream. Treating them in isolation misses the point.

We should also be upfront about something practical: dedicated GPUs are still out of reach for many practitioners, students, and independent researchers. A cookbook that requires an A100 cluster is not practical. That is why we carefully selected every model, configuration, and optimization strategy in this book to run on **free-tier Google Colab GPUs**. Every recipe has been tested in that environment. Where a technique benefits from more compute, we say so, but the default path always works on accessible hardware, whether you are in a university lab, on a coffee shop laptop, or at a personal machine with a single consumer GPU.

Finally, a recipe you cannot adapt to a new situation is a recipe you do not truly own. Every technique in this book comes with a thorough explanation of the underlying principle, not just the API call. When we introduce quantization, we explain what happens to weight distributions at different bit widths and why certain layers are more sensitive. When we cover instruction fine-tuning, we walk through the mechanics of loss masking and the dataset format choices that determine whether your model becomes a capable assistant or a confused parrot. Many LLM books end at the Jupyter notebook. This one takes you all the way to a 4-bit quantized model running behind a production endpoint with FastAPI, Docker, and high-throughput inference engines.

Our guiding principle: Show it visually. Explain it clearly. Then build it in code. In that order, on the hardware you actually have.

Who This Book Is For

This book is for ML engineers, data scientists, developers, and applied researchers with a working knowledge of Python and a foundational grasp of machine learning. If you can train a classifier and understand the gradient descent conceptually, you have everything you need. By the time you finish, you will be able to

- **Navigate the model landscape** and choose between open source models (Llama, Mistral, Gemma) and proprietary APIs (GPT, Claude, Gemini) based on your project constraints.

- **Fine-tune efficiently** with LoRA and QLoRA on consumer GPUs or free Colab instances, understanding the mathematical intuition behind adapter-based training.

- **Align models to follow instructions** through structured datasets, loss masking, and supervised fine-tuning that turns a base model into a useful assistant.

- **Quantize and optimize** with GPTQ, AWQ, and bitsandbytes, understanding which layers to protect and how bit width reduction affects quality.

- **Deploy to production** with FastAPI, Docker, and high-throughput engines like vLLM, TGI, and SGLang, applying speculative decoding and tensor parallelism.

- **Evaluate rigorously** with automated metrics, perplexity, and benchmarks like MMLU, catching regressions before users do.

- **Build beyond text** with retrieval augmented generation, vision language models, and agentic reasoning frameworks.

The book is structured as a progression but designed to be modular. If you are comfortable with Transformers and tokenization, skim Parts 1 and 2 and jump to Chapter 5. If you are newer to the field, the early chapters lay a visual and conceptual foundation that pays off throughout. For each chapter, we recommend: (1) start with the figures to build a spatial mental model, (2) read the explanation for design decisions and trade-offs, (3) run the companion Jupyter notebook on GitHub for additional experiments and edge cases, and (4) adapt the code to a different dataset, model, or deployment target. The recipes are starting points, not endpoints.

Book Structure

The book is organized into four parts that follow the natural LLM development workflow, and it concludes with Part 5 consisting of three chapters on advanced topics.

Part 1: Foundations and Environment Setup

Chapter 1 introduces the LLM development life cycle and traces the field from the Turing Test through the Transformer revolution to the current era of open access. Core building blocks like attention mechanisms, tokenization, and positional encoding are explained through annotated diagrams before any code appears. **Chapter 2** gets your environment ready, whether on Google Colab or a local machine, covering hardware fundamentals across CPUs, GPUs, and TPUs and introducing uv for fast, reproducible Python environment management.

Part 2: Models and Data Preparation

Chapter 3 surveys the model landscape, comparing open source options such as Llama, Mistral, and Gemma with proprietary APIs such as GPT, Claude, and Gemini. It walks through architectural families (encoder only, decoder only, encoder decoder) and the practical trade-offs behind model selection. **Chapter 4** covers the full data preparation pipeline: sourcing, cleaning through deduplication and PII removal, and the critical process of tokenization, with visual walkthroughs showing how raw text becomes numerical input.

Part 3: Core Techniques for Prompting and Fine-Tuning

This is the heart of the book. **Chapter 5** explores prompt engineering from zero-shot and few-shot approaches to Chain-of-Thought reasoning and ReAct. **Chapter 6** tackles full fine-tuning with transfer learning, including an honest discussion of compute cost, overfitting, and catastrophic forgetting. **Chapter 7** moves to instruction fine-tuning, showing how structured datasets and loss masking transform a base model into a capable assistant. **Chapter 8** introduces LoRA, QLoRA, and adapter methods, demonstrating how to achieve strong results by training a small fraction of parameters on a free Colab GPU. **Chapter 9** addresses data scarcity by using teacher models to generate synthetic training data from documents and domain corpora.

Part 4: Optimization, Serving, and Evaluation

Chapter 10 covers quantization in depth: INT8, 4-bit, GPTQ, AWQ, and bitsandbytes, explaining what happens to weight distributions at lower bit widths and how to navigate the accuracy-vs.-latency trade-off. **Chapter 11** takes you from static weights to scalable services using FastAPI and Docker, benchmarks vLLM, TGI, and SGLang and applies speculative decoding and tensor parallelism. **Chapter 12** builds an evaluation discipline around ROUGE, BLEU, perplexity, and MMLU, giving you the tools to catch regressions before your users do.

Part 5: Advanced Applications and Future Directions

Chapter 13 implements retrieval-augmented generation to ground models in external and private data, thereby addressing the knowledge cutoff problem. **Chapter 14** moves beyond text to vision-language models that process images and text together. **Chapter 15** explores autonomous agents built on ReAct, efficient architectures like Mixture of Experts, and the principles of responsible, ethical AI deployment.

From your first inference call to a quantized model behind a production endpoint, every recipe builds on the one before it. The figures give you the mental model, the explanations give you the reasoning, and the code gives you a working system. All code is organized as Jupyter notebooks on the companion GitHub repository, ready to run, modify, and extend. This book covers what most others split across multiple volumes: the post-training stack, the optimization stack, and the deployment stack, unified in one narrative with a shared code base and a consistent visual language. That is the gap we set out to fill, and we did it on hardware anyone can access.

PART I

Foundations and Environment Setup

## CHAPTER 1

The LLM Landscape and Core Concepts

What's Cooking: Overview of the Cookbook's Recipes and Philosophy

Welcome to the LLM Cookbook! This book is designed to be your practical guide through the exciting and rapidly evolving world of Large Language Models (LLMs). Like a well-crafted cookbook, our goal is not just to provide you with recipes that work but to help you understand the underlying principles so you can create your own innovations. In these pages, you'll find step-by-step procedures for working with LLMs—from setup and initial exploration to fine-tuning, optimization, and deployment. Each "recipe" is designed to be practical, executable code that solves specific problems while teaching broader concepts. We emphasize open source tools wherever possible, ensuring that these techniques are accessible to everyone, from individual developers to large organizations.

Our philosophy is simple: practical knowledge empowers innovation. While the theory behind LLMs is fascinating (and we'll cover essential concepts), our primary focus is on getting your hands dirty with working code. Think of this book as your experienced kitchen companion, guiding you through techniques that might otherwise take months or years to discover on your own.

The Evolution of Language Models: A Brief History

Appreciating where we are today with LLMs helps us understand the journey that brought us here. The quest to build machines that understand and generate human language dates back to the earliest days of computing, with Alan Turing's famous 1950 paper introducing the "Turing Test" setting the stage for natural language processing as a field. Early language models progressed from rule-based systems to statistical approaches like n-gram models, though these were limited in their ability to generalize.

The deep learning revolution began transforming NLP around 2013 with the introduction of word embeddings such as Word2Vec and GloVe, which capture semantic relationships between words. This was followed by recurrent neural networks (RNNs) and Long Short-Term Memory (LSTM) networks, which can process sequences, though they often struggle with long-range dependencies. A pivotal milestone was reached in 2017 with the publication of "Attention Is All You Need" by Vaswani et al., introducing the **Transformer architecture**. Its ability to process sequences in parallel and effectively weigh word importance revolutionized the field.

The modern era of LLMs dawned in 2018–2019, and influential models were built on the Transformer architecture. Google's **BERT** (Bidirectional Encoder Representations from Transformers) utilized a bidirectional encoder architecture, achieving state-of-the-art results on understanding tasks after fine-tuning. Concurrently, OpenAI's **GPT** (Generative Pre-trained Transformer) series took a different path with a decoder-only architecture focused on text generation, starting with GPT-1 and scaling up to GPT-2, which hinted at the remarkable potential unlocked by increasing model size.

This led into the **Scaling Era** (2020–2022), strongly influenced by research showing that model performance often improved predictably with scale. **GPT-3**, with its massive 175 billion parameters, represented a significant leap, demonstrating emergent capabilities such as few-shot learning. Subsequently, techniques such as **Instruction Tuning** and **Reinforcement Learning from Human Feedback (RLHF)** were developed to better align models with user intent and preferences, dramatically improving usability and leading to widespread adoption, driven by systems like **ChatGPT** and its successors, such as **GPT-4**.

The most recent phase, beginning around 2023, marks the **Democratization Era**, characterized by an explosion in the availability of powerful **open source models**. Meta's **Llama** series, Mistral AI's efficient **Mistral** and **Mixtral** models, Google's **Gemma**, TII's **Falcon**, and community efforts like **BLOOM**, among others, have made state-of-the-art capabilities accessible beyond large commercial labs. This democratization has dramatically accelerated innovation, allowing researchers and developers worldwide to experiment with, customize, and deploy powerful language models, marking a new chapter in AI accessibility. The evolution of large language models is depicted in Figure 1-1.

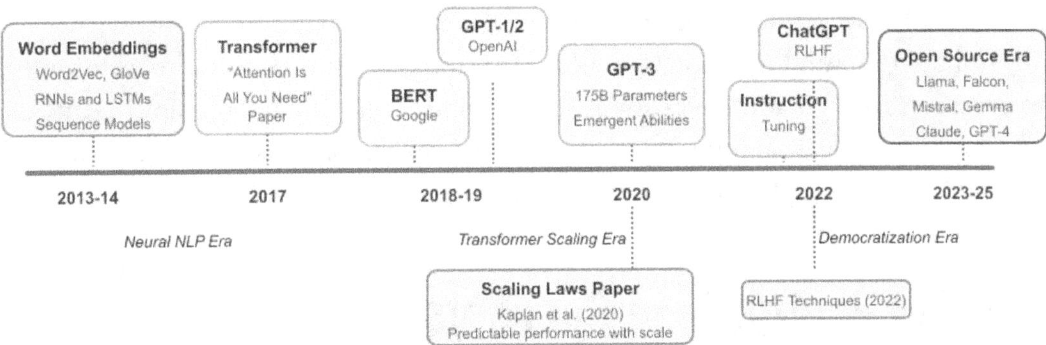

Figure 1-1. *The Evolution of Large Language Models.Timeline of LLM development across three eras: Neural NLP Era (2013-2014) with Word2Vec and RNN architectures; Transformer Scaling Era (2017-2020) featuring the "Attention Is All You Need" paper, BERT, GPT-1/2/3, and Kaplan's Scaling Laws; and Democratization Era (2022-2025) marked by RLHF in ChatGPT, instruction tuning, and open source models (Llama, Falcon, Mistral, Gemma) competing with Claude and GPT-4*

The LLM Revolution: Why LLMs Matter Today

Large Language Models have captured the public imagination and enterprise investment for good reason, they represent a fundamental shift in how we interact with and leverage artificial intelligence. Their rise is fueled by **unprecedented capabilities** that seemed like science fiction just a few years ago. Modern LLMs demonstrate remarkable abilities, such as **zero-shot or few-shot learning**, enabling them to perform tasks with minimal or no specific examples by generalizing from their vast pretraining. They exhibit **in-context learning**, adapting their behavior based on examples provided directly within the prompt. They often display **emergent abilities** like complex reasoning or code generation that only manifest at a sufficient scale. Furthermore, newer models increasingly incorporate **multimodal integration**, enabling them to process and reason about text alongside images and other data types.

These powerful capabilities are already enabling **transformative applications** across numerous fields. In **content creation**, they assist with generating articles, marketing copy, creative fiction, and even code. They power sophisticated chatbots and virtual assistants in **customer service**, offer personalized tutoring in **education**, and accelerate **software development** through code generation and debugging. LLMs are also proving valuable in **healthcare** for information summarization, in **legal services** for contract analysis, and in **scientific research** for hypothesis generation and literature review, truly reshaping how information is processed and utilized. This impact is reflected in the **ecosystem growth and investment** surrounding LLMs, with billions flowing into generative AI, specialized infrastructure emerging, and a rapidly expanding toolkit and community driving further innovation.

The LLM Pantry: Core Concepts

To work effectively with LLMs, we need to grasp several foundational concepts. The **Transformer architecture**, as mentioned, is the backbone, relying heavily on **Self-Attention Mechanisms**. These allow the model to intelligently weigh the importance of different words relative to one another within a sequence, effectively capturing context. Models don't process raw text; instead, input is converted into **Tokens**—discrete units representing words or sub-words—via **Tokenization** (using algorithms such as BPE or SentencePiece). Each token is then mapped to a dense vector representation known as an **Embedding**. These embeddings reside in a high-dimensional space in which proximity encodes semantic relationships learned during pre-training, forming the basis for transfer learning. The **Attention** mechanism often uses a Query, Key, and Value paradigm, allowing the model to determine how much attention to allocate to different input parts when producing each output (Figure 1-2). Understanding these concepts— Transformers, Attention, Tokens, and Embeddings—helps demystify model behavior and informs many practical techniques covered later.

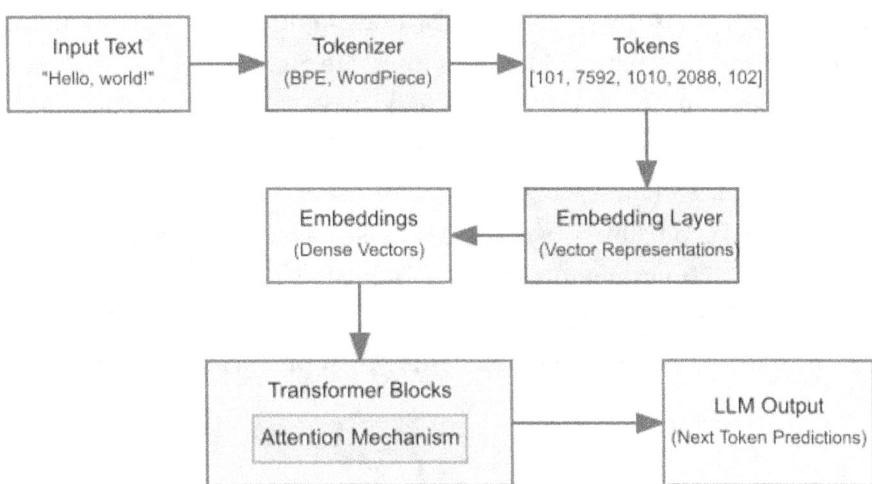

Figure 1-2. *LLM Core Concepts Flow Description from the Original Document. This diagram illustrates the core process of a Large Language Model, starting when "Input Text" is fed into a "Tokenizer" that converts it into numerical "Tokens." These tokens are then passed through an "Embedding Layer" to create "Embeddings," which are dense vector representations. Finally, these embeddings are processed by "Transformer Blocks" using an "Attention Mechanism" to generate the final "LLM Output" as token predictions*

Key Tools and Libraries Overview

Successfully navigating the LLM kitchen requires a specialized toolkit. The **Hugging Face ecosystem** has become central, providing many essential instruments. Their transformers library offers unified access to thousands of pre-trained models and tokenizers. The datasets library provides efficient tools for data handling, while tokenizers offer fast implementations of tokenization algorithms. For assessing model quality, evaluate provides standardized metrics. Crucially for this cookbook, the peft library implements Parameter-Efficient Fine-Tuning methods, and optimum assists with model optimization for deployment.

Underpinning much of this ecosystem are foundational **Deep Learning Frameworks**, with **PyTorch** being the most popular choice for LLM research and development, though TensorFlow and JAX also play significant roles. For building complex applications that chain LLM calls or integrate them with external data, frameworks such as **LangChain** and **LlamaIndex** are widely used. As efficiency is often paramount, specialized tools like bitsandbytes enable model quantization, **vLLM** or **TGI** offer optimized inference serving, and **UnslothAI** provides accelerated fine-tuning capabilities. Finally, keeping track of experiments is vital, and platforms like **Weights & Biases (W&B)** or **MLflow** are indispensable for logging metrics, visualizing results, and ensuring reproducibility in our culinary creations. We'll introduce these tools contextually throughout the book, demonstrating their practical application in LLM workflows.

From Theory to Practice: How to Use This Cookbook

The recipes in this cookbook follow a consistent pattern, outlining the objective, necessary ingredients (libraries, data, and resources), preparation steps, detailed code instructions with explanations, potential variations, and troubleshooting tips for common issues. We recommend working through the recipes sequentially initially, as concepts often build on one another. However, each recipe is also designed to be relatively self-contained, allowing you to jump to specific techniques as needed. Code examples prioritize readability and explanatory value, presented clearly with comments, ready for you to run or adapt. Figure 1-3 illustrates how the chapters flow in the book, mimicking a typical recipe flow during the preparation of a dish.

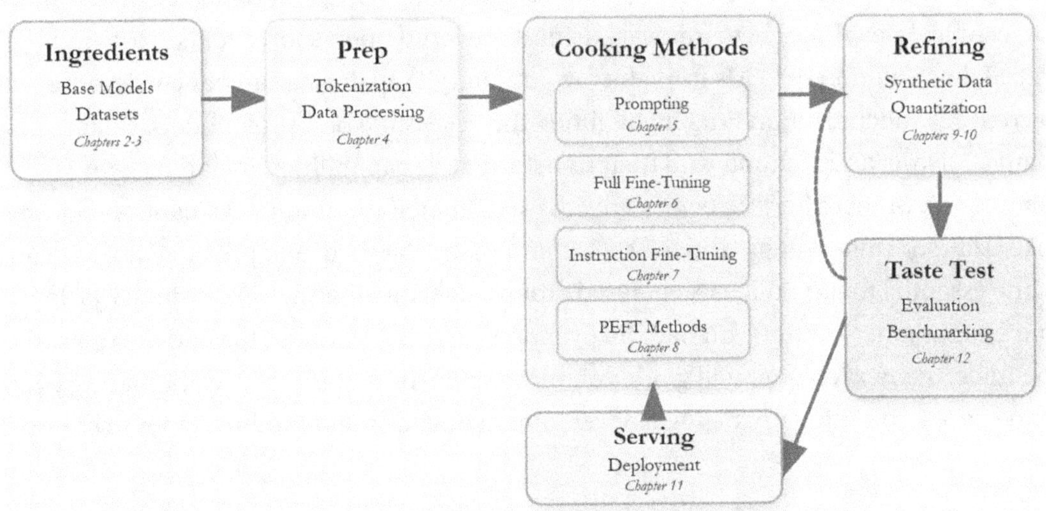

Figure 1-3. *The LLM Cookbook: From Ingredients to Serving. This flowchart visualizes the Large Language Model development process as a culinary journey. It begins with sourcing* Ingredients *(models and data) and* Prep *(data processing), followed by core* Cooking Methods *like prompting and fine-tuning. The model then enters an iterative loop of* Refining *(e.g., quantization) and* Taste Test *(evaluation) before the final* Serving *stage (deployment)*

The State of LLMs: A Rapidly Evolving Landscape

As you embark on your LLM journey, it's important to acknowledge that this field is evolving at a breathtaking pace. Research breakthroughs emerge weekly, new models are released regularly, and tools evolve rapidly. This cookbook focuses on foundational techniques expected to remain relevant but encourages continuous learning through community engagement, research papers, and experimentation. The following recipes aim to equip you with practical skills and an understanding of the underlying principles, enabling you to adapt to new developments and continue to leverage the power of LLMs effectively.

Chapter Summary

This chapter introduced the LLM Cookbook's practical, hands-on philosophy, designed for technical practitioners to innovate using a powerful open source toolkit. We traced the evolution of LLMs from the revolutionary Transformer architecture to the current "Democratization Era," highlighting the accessible power of models, such as Llama, Mistral, and Gemma, and their transformative capabilities, including zero-shot learning. To prepare for the recipes ahead, we defined foundational concepts, including tokenization, embeddings, and the self-attention mechanism, and provided an overview of the essential tools you'll use, centered around the Hugging Face ecosystem, PyTorch, and LangChain. Now that you have an overview of the LLM landscape and the tools of the trade, the next chapter will guide you through the practical first step: preparing your workspace and setting up the ideal development environment for the recipes to come.

## CHAPTER 2

Configuring Your Development Environment

Following the high-level overview of the Large Language Model (LLM) landscape, it is now time to configure your workspace. A well-organized and correctly configured development environment is essential for an efficient workflow. Proper setup from the outset saves significant time and prevents common issues, allowing you to focus on the core tasks of fine-tuning and experimenting with models. This chapter provides step-by-step instructions for setting up the most common and effective environments for LLM development, covering procedures from a quick start in a cloud-based notebook to a detailed local machine configuration and the use of professional-grade cloud GPUs.

Hardware Fundamentals: Understanding Compute Resources

LLM development relies on specific hardware components to handle the intense computational load:

- **CPU (Central Processing Unit):** A general-purpose processor essential for running your computer's operating system and handling various tasks. While a CPU can run small models or perform basic inference, it is too slow and inefficient for training.

- **GPU (Graphics Processing Unit):** The standard for serious LLM development. GPUs contain thousands of specialized cores designed to perform many calculations in parallel. This architecture is perfectly suited for the matrix multiplication operations that form the backbone of deep learning, making GPUs the primary tool for training and fine-tuning LLMs.

© Bharath Kumar Bolla, Kalpa Subbaiah and Sashi Kiran Kaata 2026
B. K. Bolla et al., *Large Language Model Recipes*, https://doi.org/10.1007/979-8-8688-2607-8_2

- The advantages of using a GPU extend beyond just raw parallel processing. Modern GPUs benefit from **high memory bandwidth**, allowing them to rapidly access the massive model weights and datasets required for training. Furthermore, they often include specialized hardware, such as NVIDIA's Tensor Cores, designed to accelerate matrix multiplication operations at the heart of deep learning. This combination of massive parallelism, high-speed memory, and specialized hardware, all supported by mature software ecosystems like NVIDIA's CUDA, is what makes GPUs not just a good choice but the essential tool for modern LLM development. Figure 2-1 depicts the GPU advantages over CPU.

- **TPU (Tensor Processing Unit):** Highly specialized hardware developed by Google. TPUs are optimized for deep learning workloads and can deliver exceptional performance, particularly at scale.

For the purposes of this guide, we will primarily focus on **GPU-based setups**, as they offer the best balance of accessibility, power, and flexibility for most developers.

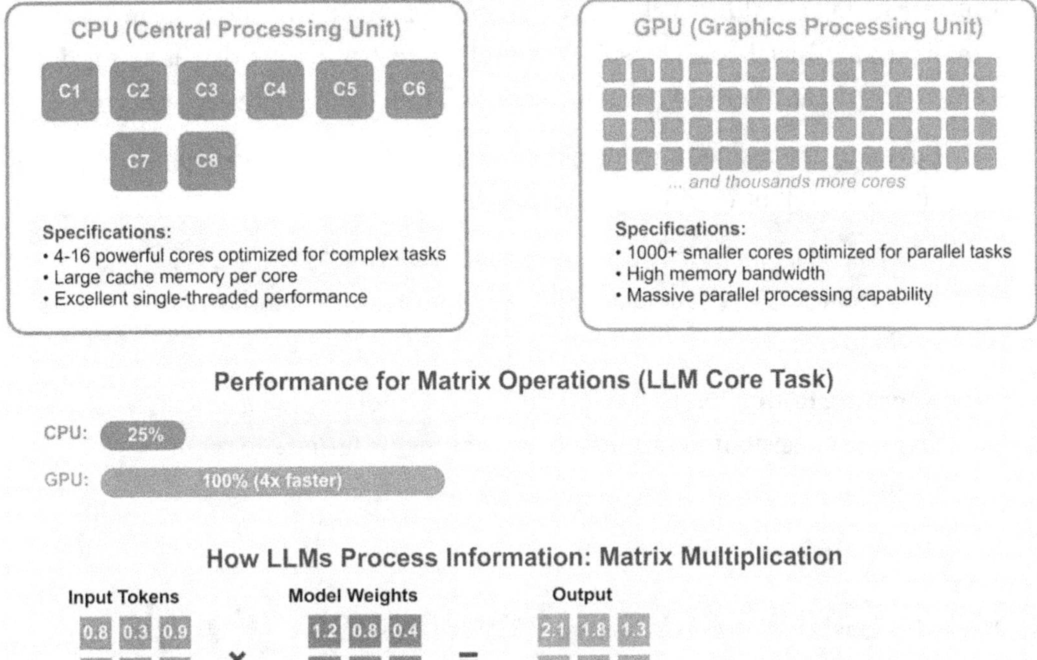

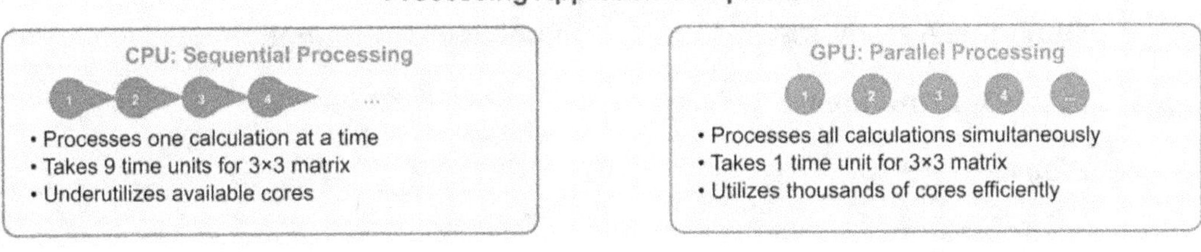

Figure 2-1. *GPU vs. CPU Architecture for Large Language Model Processing. Comparison of CPU (8 large cores, sequential processing) and GPU (thousands of small cores, parallel processing) architectures. Performance metrics show 4x GPU speed advantage for matrix operations essential to LLMs. Matrix multiplication example demonstrates parallel computation of input tokens × model weights = output*

Recommended Tool: uv for Package Management

To install and manage Python packages, we recommend using uv. It is an extremely fast Python package installer and resolver, written in Rust, and intended as a drop-in replacement for pip and venv. We recommend uv for several key advantages:

- **Speed:** It can resolve and install packages 5-10× faster than pip, reducing wait times during setup and iteration.

- **Efficiency:** uv includes a high-performance virtual environment manager (uv venv), replacing the need for Python's standard venv or Conda.

- **Reproducibility:** Its speed and modern dependency resolver make creating consistent, reproducible environments faster and more reliable.

Guide 1: Cloud Notebook Environment (Google Colab Quickstart)

Google Colab provides a free, cloud-based environment that includes GPU access. It is an excellent way to begin immediately with zero initial setup cost:

- **Outcome:** A ready-to-use LLM development environment.

- **Time:** 5–10 minutes

- **Steps:**

    1. **Open a New Notebook:** Navigate to colab.research.google.com and create a new notebook.

    2. **Enable GPU Acceleration:** Go to Runtime ➤ Change runtime type and select GPU.

3. **Verify GPU Allocation:** Run `!nvidia-smi` in a cell to confirm GPU activation and view its specifications.

    Install Essential Libraries with uv:

    ```
    !pip install -q uv
    ```

    ```
    !uv pip install -q -U transformers datasets accelerate peft
    bitsandbytes wandb
    ```

4. **Authenticate Services:** Use `notebook_login()` for Hugging Face and `wandb.login()` for Weights & Biases to connect your accounts.

Guide 2: Local Machine Setup with uv

A local environment is the optimal choice for developers who require full control and a persistent setup. This guide uses uv to create a fast, modern, and self-contained environment:

- **Outcome:** A powerful, persistent LLM environment.

- **Time:** 15–30 minutes

- **Prerequisites:** An NVIDIA GPU with at least 8GB of VRAM and the latest NVIDIA drivers.

- **Steps:**

    1. **Install uv:** Run `pip install uv` in your terminal.

    2. **Create and Activate Workspace:** Run `uv venv` to create a virtual environment and then activate it.

    3. **Install PyTorch with CUDA:** Use the official `pip` command from the PyTorch website, but execute it with `uv pip`.

    4. **Install Core Libraries:** Run `uv pip install -U transformers datasets accelerate peft bitsandbytes wandb`.

    5. **Authenticate Services:** Use `huggingface-cli login` and `wandb login` in your terminal.

Guide 3: Using Specialist Cloud GPU Providers

If a local machine is not powerful enough and Colab is too limited, a dedicated cloud GPU provider such as RunPod, Lambda Labs, or Vast.ai is a viable alternative. These services offer a user-friendly and often more affordable path to high-performance computing:

- **Advantages:** They offer simplicity, cost-effectiveness, and access to state-of-the-art hardware without the complexity of major cloud platforms.

- **General Workflow:**

    1. **Create an Account and Launch an Instance:** Sign up and launch a virtual machine with your desired GPU and a pre-installed machine learning template.

    2. **Connect to the Instance:** Connect to the machine via SSH.

    3. **Set Up Your uv Environment:** The process is nearly identical to Guide 2. It is critical to create your environment in a persistent storage volume to ensure your setup survives instance reboots.

Key Takeaways

This chapter provided a practical guide to setting up an LLM development environment, ensuring a solid foundation for subsequent procedures:

- **Hardware Is a Critical Factor:** GPUs are the standard for LLM development due to their massive parallel processing capabilities, high memory bandwidth, and specialized hardware like Tensor Cores, which are essential for the matrix mathematics that power these models.

- **Modern Tooling Improves Efficiency:** Using a fast package manager like uv streamlines setup by accelerating package installation and simplifying virtual environment management.

- **Select the Appropriate Environment for the Task:** We covered three primary paths for configuring a workspace:

  - **Google Colab:** Best for quick, free-of-charge experiments and learning.

  - **Local Machine:** Ideal for developers who require full control and a persistent environment for ongoing projects.

  - **Specialist Cloud GPU:** The optimal choice for serious, high-performance work, offering access to state-of-the-art hardware with greater simplicity and cost-effectiveness than major cloud providers.

With your development environment configured, you are prepared to work with the core components of LLM systems: the models themselves. The next chapter will explore the vast range of available LLMs, from powerful open source options to state-of-the-art proprietary APIs.

Chapter Summary

In this chapter, we provided a practical guide to configuring an efficient and persistent Large Language Model (LLM) development environment. We established that hardware is a critical factor, with the GPU the standard due to its massive parallel processing capabilities, high memory bandwidth, and specialized Tensor Cores, which are well suited to deep learning's matrix operations. We recommended modern tooling, specifically uv, as an extremely fast package installer and resolver to streamline setup by accelerating package installation and simplifying virtual environment management. Finally, we detailed three distinct, recommended paths for configuring a workspace: Google Colab for quick, free-of-charge experiments; a Local Machine for full control and persistent projects; and Specialist Cloud GPU Providers for serious, high-performance work, offering simplicity and cost-effectiveness.

With your development environment configured, you are prepared to work with the core components of LLM systems. The next chapter will explore the vast range of available LLMs, from powerful open source options to state-of-the-art proprietary APIs.

PART II

Models and Data Preparation

## CHAPTER 3

LLM Architectures and Foundational Models

With the development environment configured, our focus now shifts to the core component of any LLM system: the model itself. The selection of a base model is a critical decision that dictates the functional capabilities and performance ceiling of the final application. The current landscape of Large Language Models is diverse, with models categorized by several key characteristics. In this chapter, we will conduct a systematic survey of this landscape. We will begin by examining the primary architectural paradigms that grant models their specialized abilities and then analyze the strategic trade-offs between using community-driven open source models and powerful proprietary APIs. The model selection can be broadly put into two phases.

Phase 1: Model Selection Strategy: Open Source vs. Proprietary APIs

The first strategic decision in an LLM project is choosing the model's source and access model. This choice determines the level of control, cost, and maintenance required. Generally, it falls into two primary categories: leveraging community-driven open source models or using highly managed proprietary models accessed via APIs.

Option 1: The Open Source Ecosystem (Transparency and Control)

This category consists of models where the weights, architecture, and often training code are publicly released. Influential examples include Meta's Llama series, Mistral AI's Mistral/Mixtral models, and Google's Gemma family. The defining characteristic is

© Bharath Kumar Bolla, Kalpa Subbaiah and Sashi Kiran Kaata 2026
B. K. Bolla et al., *Large Language Model Recipes*, https://doi.org/10.1007/979-8-8688-2607-8_3

access; developers can download, inspect, modify, and run these models on their own hardware. This approach is centered around community platforms like the Hugging Face Hub.

Key Benefits and Challenges of Open Source LLMs

Advantages	Disadvantages
Transparency & Control (Weights/Arch)	Resource Intensive (Compute/VRAM)
Customization & Deep Fine-tuning	Setup & Maintenance Overhead
No Per-Use API Costs (Compute costs apply)	Potentially Less Polished/Aligned (Out of the Box)
Enhanced Data Privacy (Local processing)	Community/Internal Support Reliance
Active Community & Rapid Innovation	License Restrictions (Varying)

The following procedure details the standard method for accessing and loading these open source models.

Recipe 1: Loading an Open Source Model

Accessing models from the Hugging Face Hub

Goal: Load a pre-trained open source model and its tokenizer from the Hugging Face Hub.

What This Code Does (Recipe 1)

- Loads a chosen open source model name (e.g., Gemma, Mistral) from the Hugging Face Hub

- Initializes the tokenizer responsible for converting text into model-readable tokens

- Loads the model weights with automatic device placement (CPU/GPU) and efficient precision settings

- Uses modular helper functions, making the same loading utilities reusable across other recipes

- Performs a small test generation to confirm that the model and tokenizer work together correctly

```python
# --- Recipe: Loading an Open Source Model from Hugging Face Hub ---
# Goal: Load a pre-trained open-source model and its tokenizer.
# Library: Hugging Face Transformers
# Note: Ensure you have run `pip install transformers torch accelerate` in
  your environment.
#        Some models (like Meta's Llama) are "gated" and require
          authentication.

from transformers import AutoTokenizer, AutoModelForCausalLM
import torch
import os

# --- Configuration ---
# Choose a model ID from the Hugging Face Hub.
# Examples:
# 'mistralai/Mistral-7B-Instruct-v0.1' # Good performance, Apache
  2.0 license
# 'google/gemma-2b-it' # Google's Gemma instruct-tuned model
# 'distilgpt2' # Very small, good for quick tests
model_id = "google/gemma-2b-it" #Using Gemma 2B instruct as an example

from huggingface_hub import HfApi, login

# Set HF Token as ENV
os.environ["HF_TOKEN"] = "hf_xxxxxxxxxxxxxxxxxxxxxx" #replace with
your token

#Use the ENV variable
hf_token = os.getenv("HF_TOKEN")
if not hf_token:
    raise ValueError("HF_TOKEN environment variable not set")
# Optional: login once (recommended)
login(token=hf_token)
# Verify identity
api = HfApi()
whoami = api.whoami(token=hf_token)
print(whoami)

    # --- Authentication (Optional - Needed for Gated Models like Llama) ---
```

```python
# If using a gated model, you need to:
# 1. Accept the license terms on the model's Hugging Face page.
# 2. Log in using the Hugging Face CLI: `huggingface-cli login`
#    This saves your token locally. The library will then use it
#      automatically.
# Make sure you have given the read permission while creating the token.

# Alternatively, provide the token explicitly (less secure):
#use_auth_token = 'REPLACE_WITH_YOUR_HUGGING_FACE_TOKEN'

use_auth_token = hf_token
# Set to None or False if model is not gated or logged in via CLI
print(f"Loading tokenizer for model: {model_id}")
try:
    # 1. Load the Tokenizer
    #    AutoTokenizer automatically selects the correct tokenizer class
    #      based on the model ID.
    tokenizer = AutoTokenizer.from_pretrained(model_id, use_auth_token=use_
    auth_token)
    print("Tokenizer loaded successfully.")
except Exception as e:
    print(f"Error loading tokenizer: {e}")
    print("Check model ID, internet connection, and authentication if
    required.")
    exit()

print(f"\nLoading model: {model_id}")
print("This might take a while depending on model size and download
speed...")
try:
    # 2. Load the Model
    #    AutoModelForCausalLM is suitable for text generation models.
    #    Use `device_map='auto'` to automatically distribute the model
    #      across available GPUs (requires accelerate).
    #    Use `torch_dtype=torch.bfloat16` (if supported by GPU) or torch.
    #      float16 for memory savings.
```

```python
    model = AutoModelForCausalLM.from_pretrained(
        model_id,
        token=use_auth_token,
        device_map="auto", # Automatically use available GPU(s) or CPU
        torch_dtype=torch.bfloat16 # Use bfloat16 for efficiency if
        available
    )
    print(f"Model loaded successfully onto device: {model.device}") # Will
    show cuda:0 if GPU is used
except Exception as e:
    print(f"Error loading model: {e}")
    print("Check model ID, internet connection, authentication, and
    available GPU memory.")
    exit()

# --- Verification ---
# You now have the 'tokenizer' and 'model' objects ready for use in other
recipes.
print("\nModel and Tokenizer are ready!")

# Example: Tokenize a sample text
sample_text = "Hello, LLM Chef!"
tokens = tokenizer.encode(sample_text)
print(f"\nSample text: '{sample_text}'")
print(f"Tokens: {tokens}")
decoded_text = tokenizer.decode(tokens)
print(f"Decoded tokens: '{decoded_text}'")
```

Option 2: The Proprietary Model Landscape (Performance and Convenience)

This category features state-of-the-art models developed and hosted by commercial entities like OpenAI (GPT series), Anthropic (Claude series), and Google (Gemini series). They are accessed via a paid **Application Programming Interface (API)**. The underlying architecture and weights are proprietary, presenting a powerful but opaque "black box."

Advantages vs. Limitations of Commercial LLM Services

Advantages	Disadvantages
State-of-the-Art Performance (Often)	Cost (Pay-per-Use API Calls)
Ease of Use & Managed Infrastructure	Lack of Control/Transparency ("Black Box")
High Availability & Scalability	Data Privacy Concerns (Data sent to vendor)
Strong Safety/Alignment Measures (Often)	Vendor Lock-In & Limited Switching
Advanced Features (e.g., Function Calling)	Limited Customization/Fine-Tuning Options

The following procedure demonstrates how to interact with these models programmatically.

Recipe 2: Calling a Proprietary Model API

Interacting with models via vendor APIs

Goal: Demonstrate making an API call to a proprietary LLM endpoint.

What This Code Does (Recipe 2)

- Initializes a client for a proprietary LLM service (such as OpenAI)

- Sends a prompt request to the model using its official SDK

- Receives and prints the model's generated response along with metadata such as token usage

- Shows an alternative way to call a model using a generic HTTP request instead of a vendor-specific SDK

- Processes the HTTP response to extract and display the generated text

```python
# --- Recipe: Calling a Proprietary Model API ---
# Goal: Demonstrate making an API call to a hypothetical closed-source LLM
endpoint.
# Library: requests (standard Python library), openai (example specific
library)
# Note: This uses hypothetical examples. Replace with actual API details
from your provider.
#        NEVER hardcode API keys directly in code for production systems.
         Use environment variables or secrets management.

import requests
import os
import json

# --- Using a specific library like OpenAI (Recommended if available) ---
# Ensure you have the library installed: pip install openai
# Requires setting the API key as an environment variable:
# export OPENAI_API_KEY='sk-YOUR_ACTUAL_API_KEY'

# Example using the OpenAI library structure (adapt for other providers
like Anthropic, Google)
print("--- Example using OpenAI library ---")

try:
    from openai import OpenAI
    # The client automatically picks up the OPENAI_API_KEY environment
      variable
    client = OpenAI() # Add base_url if using a non-OpenAI compatible
    endpoint

    # Check if API key is set (optional but good practice)
    if not client.api_key:
        print("Error: OPENAI_API_KEY environment variable not set.")
    else:
        print("OpenAI client initialized.")
        prompt_text = "Explain the concept of Parameter-Efficient Fine-
        Tuning (PEFT) in one paragraph."
        model_to_use = "gpt-3.5-turbo" # Or other available model
        like "gpt-4"
```

```python
        print(f"\nSending prompt to model: {model_to_use}")
        completion = client.chat.completions.create(
            model=model_to_use,
            messages=[
                {"role": "system", "content": "You are a helpful AI
                    assistant explaining complex ML concepts simply."},
                {"role": "user", "content": prompt_text}
            ],
            max_tokens=150, # Limit the length of the response
            temperature=0.7 # Controls randomness (0=deterministic,
            >1=more random)
        )

        # Extract and print the response
        response_text = completion.choices[0].message.content
        print("\nAPI Response:")
        print(response_text)
        print("\nUsage Info:")
        print(completion.usage) # Shows token usage

except ImportError:
    print("OpenAI library not found. Skipping OpenAI example.")
except Exception as e:
    print(f"Error during OpenAI API call: {e}")

#API CALL
# --- Generic Example using 'requests' library ---
# Useful if the provider doesn't have a dedicated Python library or for
simple calls.
print("\n--- Example using generic 'requests' library ---")

# --- Configuration (Replace with actual values) ---
# **NEVER COMMIT ACTUAL KEYS TO VERSION CONTROL**
# Load from environment variables is best practice:
# api_key = os.environ.get("VENDOR_API_KEY")
api_key = os.getenv("OPENAI_API_KEY") # Replace or load from env var
api_endpoint_url = "https://api.openai.com/v1/chat/completions"
# Replace with actual endpoint
```

```python
if api_key == "YOUR_VENDOR_API_KEY_PLACEHOLDER":
    print("Warning: Using placeholder API key. Set a real key for
    actual use.")

# --- Prepare Request ---
headers = {
    "Authorization": f"Bearer {api_key}",
    "Content-Type": "application/json",
}

messages =  [
    {"role": "user", "content": "Write a short tagline for an LLM
    cookbook."}
  ]

data = {
    "model": "gpt-3.5-turbo", # Replace with actual model name
    "messages": messages,
    "max_tokens": 20,
    "temperature": 0.8
}

print(f"\nSending request to generic endpoint: {api_endpoint_url}")
try:
    # --- Make API Call ---
    response = requests.post(api_endpoint_url, headers=headers, data=json.
    dumps(data))
    response.raise_for_status() # Raise an exception for bad status codes
    (4xx or 5xx)

    # --- Process Response ---
    response_data = response.json()
    print("\nAPI Response (JSON):")
    print(json.dumps(response_data, indent=2))

    # Extract the actual generated text (structure depends on API provider)
    # This is a hypothetical structure, adjust based on the actual API
      response
```

```python
    if "choices" in response_data and len(response_data["choices"]) > 0:
        generated_text = response_data["choices"][0]["message"].
        get("content", "N/A")
        print(f"\nGenerated Text: {generated_text.strip()}")
    else:
        print("\nCould not extract generated text from response.")

except requests.exceptions.RequestException as e:
    print(f"Error during generic API call: {e}")
    if response is not None:
        print(f"Response status code: {response.status_code}")
        print(f"Response text: {response.text}")
except Exception as e:
    print(f"An unexpected error occurred: {e}")

# --- End of Recipe -
```

Phase 2: A Technical Guide to the Three Core Architectures

Beyond the crucial distinction between open source access and proprietary APIs, another fundamental way to understand the models on our buffet is by their core **Transformer architecture**. While most modern LLMs leverage the Transformer, variations in how its components (encoders and decoders) are used, and how they are pre-trained, lead to distinct strengths and weaknesses (Figure 3-1). Let's briefly explore the main architectural styles and their typical training objectives.

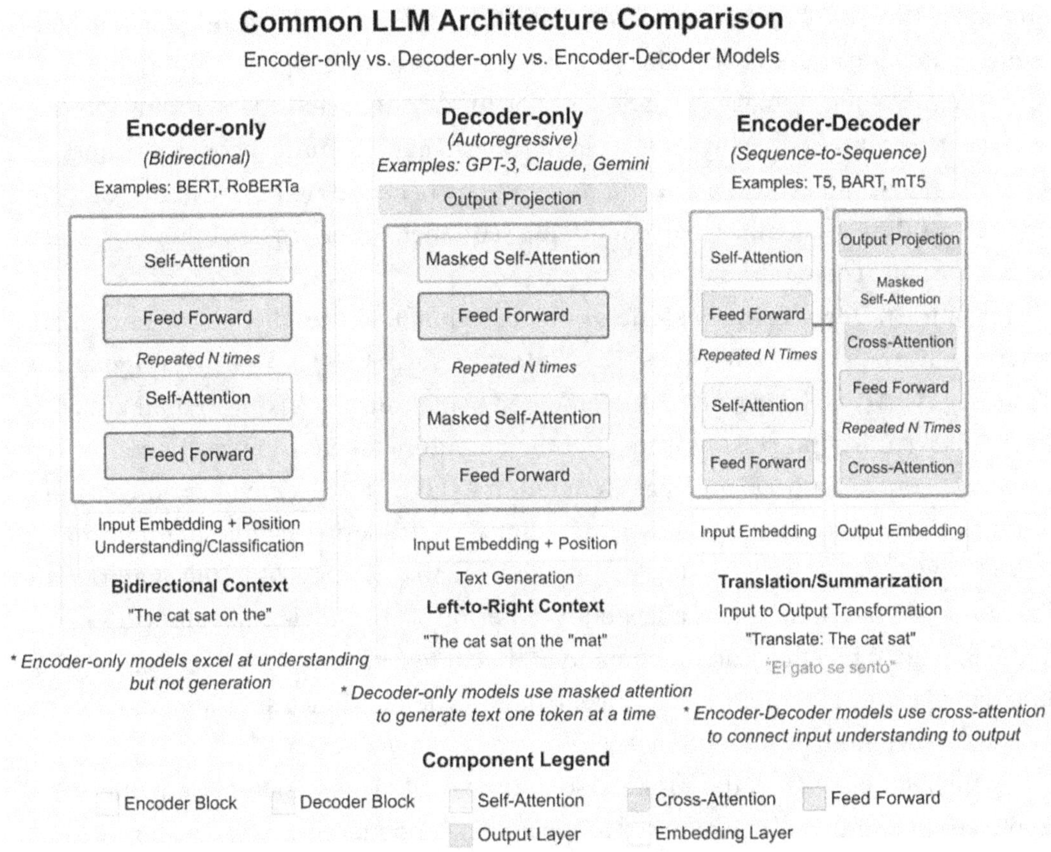

Common LLM Architecture Comparison

Encoder-only vs. Decoder-only vs. Encoder-Decoder Models

Figure 3-1. *This figure illustrates the three fundamental LLM architecture types: Encoder-only (BERT, RoBERTa), Decoder-only (autoregressive models), and Encoder-Decoder (T5, BART, mT5). Each architecture features distinct structural components and specialized capabilities. Encoder-only models employ bidirectional self-attention for understanding but not generation. Decoder-only models use masked self-attention with left-to-right processing for text generation. Encoder-Decoder models combine both approaches with cross-attention mechanisms to transform input sequences into different output sequences, making them ideal for translation and summarization tasks*

First, we encounter **Encoder-Only Models**, exemplified by architectures like BERT, RoBERTa, and DistilBERT. These models are designed to build a deep understanding of the input text by processing the entire sequence at once. This allows information to flow bidirectionally, meaning each token can "attend" to all other tokens, capturing rich contextual relationships. They are typically pre-trained using a **Masked Language Modeling (MLM)** objective, where some input tokens are randomly masked (e.g., replaced with a [MASK] token), and the model learns to predict these original masked tokens by looking at the surrounding unmasked context from both left and right. Consequently, their primary strength lies in Natural Language Understanding (NLU) tasks, such as sentiment analysis, text classification, named entity recognition, and generating high-quality text embeddings. However, because their pre-training objective focuses on filling gaps rather than generating sequences, they are less inherently suited for free-form text generation.

In contrast, **Decoder-Only Models** operate sequentially, predicting the next token based solely on the preceding ones. This **Autoregressive (AR)** or **Causal Language Modeling (CLM)** objective, often implemented using causal masking in their attention mechanism (preventing tokens from attending to future tokens), makes them naturally adept at Natural Language Generation (NLG). This architecture powers the vast majority of recent popular LLMs, including the GPT series, Llama, Gemma, and Mistral. They excel at tasks like text completion, creative writing, building chatbots, and dialogue systems. Their remarkable ability to perform few-shot learning is also a key characteristic. While they can understand input context effectively through prompting, their core design isn't primarily optimized for tasks demanding a simultaneous, deep analysis of the entire input sequence in the same way encoders are.

Finally, **Encoder-Decoder Models**, sometimes called Sequence-to-Sequence (Seq2Seq) models, combine both components. The encoder processes the entire input sequence to create a contextual representation (similar in spirit to MLM's goal of understanding context), while the decoder generates the output sequence auto-regressively (like CLM), conditioned on the encoder's output. Their pre-training often involves specialized objectives suited for sequence-to-sequence tasks, like T5's span corruption (a variation of MLM adapted for Seq2Seq). This architecture is particularly well suited for tasks where the output is a direct transformation of the input, such as machine translation, abstractive summarization, and certain styles of question answering. Prominent examples include T5 and BART. While powerful for these transformation tasks, they can be more complex than decoder-only models and might not match the zero-shot generative prowess of the largest decoder-only architectures.

***Table 3-1.** Comparing LLM Architectures: Design, Training, and Use Cases*

Architecture	How It Works (Simplified)	Pre-training Objective	Primary Use Cases	Examples
Encoder-Only	Processes full input sequence at once (bidirectional context)	Masked LM (MLM)	Natural Language Understanding (NLU): Classification, NER, Embeddings	BERT, RoBERTa
Decoder-Only	Predicts next token based on previous tokens (auto-regressive)	Causal LM (CLM / AR)	Natural Language Generation (NLG): Chat, Text Completion, Summarization	GPT series, Llama
Encoder-Decoder	Encoder processes input, Decoder generates output	Seq2Seq Objectives	Sequence-to-Sequence (Seq2Seq): Translation, Summarization, Q&A	T5, BART

The LLM Landscape: A Two-Dimensional Overview

By combining the two concepts of the **access model** and **architecture**, we can map the current LLM landscape. This helps in strategically positioning different models based on their core function (understanding, generation, or transformation) and their accessibility (open source or proprietary). Figure 3-2 illustrates the current LLM landscape.

Once a model is loaded or accessible via an API, the next step is to perform inference. The Hugging Face pipeline function provides a high-level abstraction for this, simplifying common tasks.

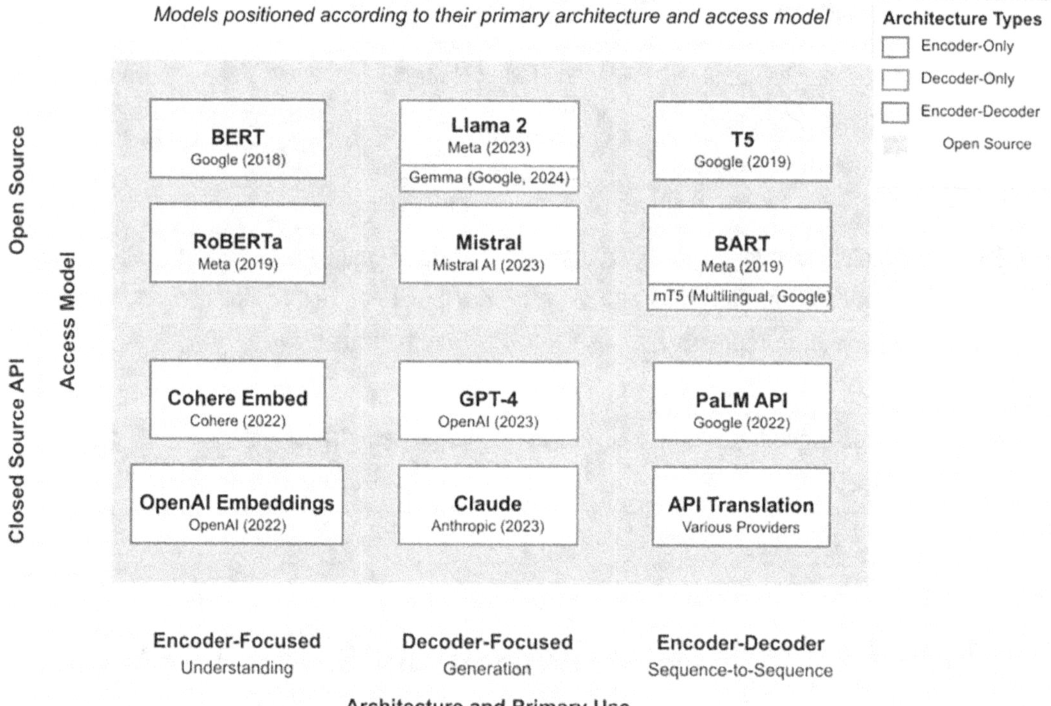

Figure 3-2. *This matrix plots twelve major LLMs across architecture types (Encoder, Decoder, and Encoder-Decoder) and access models (Open Source vs. API-only). Open source models include BERT/RoBERTa, Llama/Mistral, and T5/ BART, while closed APIs include Cohere/OpenAI embeddings, GPT-4/Claude, and translation services. Positioning indicates primary function: understanding, generation, or sequence transformation*

Recipe 3: Basic Inference Tasks

Using a loaded model for generation and zero-shot classification

Goal: Use the Hugging Face pipeline for a common task like text classification.

What This Code Does (Recipe 3)

- Uses the Hugging Face **pipeline** abstraction to perform common inference tasks

- Demonstrates **text generation** using a small decoder-only model (`distilgpt2`)

- Performs **zero-shot classification** using an NLI-based model (facebook/bart-large-mnli)

- Shows how to reuse models already loaded in memory using the utility helpers from **Recipe 1**

- Compares results between using a model ID directly and using **pre-loaded model + tokenizer** for finer control

A zero-shot classification is the process of classifying the samples without any model training. With LLMs, we can leverage their inherent capability for zero-shot classification with reasonable accuracy.

```
# --- Recipe: Basic Inference Tasks (Generation & Zero-Shot
Classification) ---
# Goal: Use a loaded model for common tasks via Hugging Face pipelines.
# Library: Hugging Face Transformers
# Prerequisite: A model and tokenizer should be loaded (e.g., from ch3_
recipe_load_oss).
#               Or use a model ID directly within the pipeline.

from transformers import pipeline, AutoTokenizer, AutoModelForCausalLM,
AutoModelForSequenceClassification
import torch

# --- Option 1: Use a model ID directly in the pipeline (Easier for
standard tasks) ---
print("--- Option 1: Using Pipelines with Model ID ---")

# Text Generation
print("\nLoading text-generation pipeline...")
try:
    # Using a small model for quick demo
    generator = pipeline('text-generation', model='distilgpt2', device=0 if
    torch.cuda.is_available() else -1) # Use GPU if available
    prompt = "The secret ingredient in the best AI recipes is"
    print(f"Generating text for prompt: '{prompt}'")
    outputs = generator(prompt, max_new_tokens=30, num_return_sequences=1)
    # max_new_tokens generates 30 tokens *after* the prompt
```

```
    print("Generated Text:")
    print(outputs[0]['generated_text'])
except Exception as e:
    print(f"Error during text generation: {e}")

# Zero-Shot Classification
print("\nLoading zero-shot-classification pipeline...")
try:
    # Using a model fine-tuned for Natural Language Inference (NLI),
      suitable for zero-shot
    classifier = pipeline("zero-shot-classification", model="facebook/bart-
    large-mnli", device=0 if torch.cuda.is_available() else -1)
    sequence_to_classify = "This cookbook focuses on practical LLM
    fine-tuning."
    candidate_labels = ["machine learning", "cooking", "finance", "sports"]
    print(f"\nClassifying sequence: '{sequence_to_classify}'")
    print(f"With candidate labels: {candidate_labels}")
    results = classifier(sequence_to_classify, candidate_labels)
    print("\nClassification Results:")
    # Print results sorted by score
    sorted_results = sorted(zip(results['labels'], results['scores']),
    key=lambda x: x[1], reverse=True)
    for label, score in sorted_results:
        print(f"- {label}: {score:.4f}")

except Exception as e:
    print(f"Error during zero-shot classification: {e}")

# --- Option 2: Use pre-loaded model and tokenizer (More control, useful if
model already loaded) ---
print("\n--- Option 2: Using Pipelines with Pre-loaded Model/
Tokenizer ---")
# This assumes you have 'model' and 'tokenizer' variables from ch3_recipe_
load_oss
# We'll reload Gemma here for demonstration if the previous recipe wasn't
run contiguously.
```

```python
model_id_loaded = "google/gemma-2b-it" # Make sure this matches the model
you intend to load/use
preloaded_model = None
preloaded_tokenizer = None

try:
    print(f"\nAttempting to load {model_id_loaded} for Option 2 demo...")
    preloaded_tokenizer = AutoTokenizer.from_pretrained(model_id_loaded)
    # Load specifically for Causal LM if planning text generation
    preloaded_model = AutoModelForCausalLM.from_pretrained(
        model_id_loaded,
        device_map="auto",
        torch_dtype=torch.bfloat16 # Use bfloat16 for efficiency if
        available
    )
    print("Pre-loaded model and tokenizer ready.")

    # Create pipeline using the loaded components
    generator_loaded = pipeline('text-generation', model=preloaded_model,
    tokenizer=preloaded_tokenizer) # device is inferred from model.device

    prompt_loaded = "To build a great AI application, you need"
    print(f"\nGenerating text using pre-loaded model: '{prompt_loaded}'")
    outputs_loaded = generator_loaded(prompt_loaded, max_new_tokens=30,
    num_return_sequences=1, do_sample=True, temperature=0.7)
    print("Generated Text (Pre-loaded):")
    print(outputs_loaded[0]['generated_text'])

    # Note: For zero-shot classification with a pre-loaded model, ensure
      the loaded model
    # is suitable for classification (e.g.,
      AutoModelForSequenceClassification) and the task.
    # Using a CausalLM like Gemma directly in a zero-shot pipeline might
      not yield optimal results
    # without specific fine-tuning or prompt engineering for that task.
    print("\n(Skipping zero-shot with pre-loaded CausalLM model as it's not
    ideal for the task without adaptation)")
```

```
except NameError:
    print("\nSkipping Option 2 as 'model' and 'tokenizer' variables are not
    defined.")
    print("(This likely means the 'ch3_recipe_load_oss' recipe wasn't run
    just before this one).")
except Exception as e:
    print(f"\nError during Option 2 setup or inference: {e}")
```

Building on this, a common and valuable exercise is to compare the outputs
of different models for the same prompt to understand their relative strengths and
weaknesses.

Recipe 4: Comparing Model Outputs

Running the same prompt through different models

Goal: Run the same prompt through two different models and compare their
outputs.

What This Code Does (Recipe 4)

- Loads **two different language models** to run the same prompt
  through both

- Uses a small model directly ("distilgpt2") and a larger model via the
  Recipe 1 utility helpers

- Generates output from each model for a common input sentence

- Prints both results side by side to observe differences in quality,
  reasoning, detail, and style

- Helps build an intuition for how model size, architecture, and tuning
  impact text generation

```
# --- Recipe: Comparing Outputs of Different Models ---
# Goal: Run the same prompt through two different models and compare their
outputs.
# Library: Hugging Face Transformers

from transformers import pipeline, set_seed
import torch
```

```python
# --- Configuration ---
# Choose two different model IDs
model_id_1 = "distilgpt2" # Smaller, faster model
model_id_2 = "google/gemma-2b-it" # Larger, potentially more capable model
(requires more resources)

prompt = "The future of artificial intelligence is"

# Set seed for reproducibility of generation (if models use sampling)
set_seed(123)

# --- Load Pipelines for Both Models ---
print(f"Loading pipeline for Model 1: {model_id_1}")
try:
    # Use device=0 for GPU if available, otherwise -1 for CPU
    device_index = 0 if torch.cuda.is_available() else -1
    generator1 = pipeline('text-generation', model=model_id_1,
    device=device_index)
    print("Pipeline 1 loaded.")
except Exception as e:
    print(f"Error loading pipeline 1: {e}")
    generator1 = None # Ensure variable exists but is None

print(f"\nLoading pipeline for Model 2: {model_id_2}")
try:
    # Gemma might require bfloat16 for efficiency on some GPUs
    dtype = torch.bfloat16 if torch.cuda.is_available() and torch.cuda.
    is_bf16_supported() else torch.float32
    generator2 = pipeline('text-generation', model=model_id_2,
    device=device_index, torch_dtype=dtype)
    print("Pipeline 2 loaded.")
 except Exception as e:
    print(f"Error loading pipeline 2: {e}")
    print("This could be due to insufficient GPU memory for Gemma-2B.")
    generator2 = None # Ensure variable exists but is None

 # --- Generate and Compare Outputs ---
 print(f"\n--- Comparing Outputs for Prompt: '{prompt}' ---")
```

```python
output1 = "Pipeline 1 failed to load."
if generator1:
    print(f"\nGenerating with Model 1 ({model_id_1})...")
    try:
        outputs1 = generator1(prompt, max_new_tokens=50, num_return_
        sequences=1, do_sample=True, temperature=0.7)
        output1 = outputs1[0]['generated_text']
    except Exception as e:
        output1 = f"Error during generation with Model 1: {e}"
print(f"\nOutput from {model_id_1}:\n{output1}")

output2 = "Pipeline 2 failed to load or generation failed."
if generator2:
    print(f"\nGenerating with Model 2 ({model_id_2})...")
    try:
        # Gemma instruct models often need specific prompt formatting if
          not using pipeline defaults
        # For basic comparison, we'll use the raw prompt here.
        outputs2 = generator2(prompt, max_new_tokens=50, num_return_
        sequences=1, do_sample=True, temperature=0.7)
        output2 = outputs2[0]['generated_text']
    except Exception as e:
        output2 = f"Error during generation with Model 2: {e}"
print(f"\nOutput from {model_id_2}:\n{output2}")

print("\n--- Comparison Complete ---")
# Observe differences in style, coherence, length, factual accuracy (if
applicable), etc.
```

Chapter Summary

In this chapter, we surveyed the modern landscape of Large Language Models along
two main axes: access model and architecture. On the access side, we contrasted
community-driven open source models (such as Llama, Mistral, and Gemma) with
proprietary API-based models (such as GPT, Claude, and Gemini), highlighting the
trade-offs between maximal control and convenience vs. state-of-the-art performance.

Architecturally, we categorized models into three Transformer families: encoder-only (e.g., BERT), which excels at Natural Language Understanding (NLU), decoder-only (e.g., GPT, Llama), which is naturally suited for Natural Language Generation (NLG), and encoder–decoder (e.g., T5, BART), which is ideal for Sequence-to-Sequence (Seq2Seq) transformation tasks.

We then translated these concepts into practice through four recipes, demonstrating how to load open source models from the Hugging Face Hub, call proprietary models via vendor APIs, perform basic inference tasks such as generation and zero-shot classification using the Hugging Face pipeline, and compare outputs across different models. Armed with a clearer map of the LLM landscape and practical recipes for loading, calling, and comparing models, the next essential step is to prepare the data that will be processed. The following chapter will address the principles and techniques of data curation, cleaning, and tokenization.

## CHAPTER 4

Data Preparation and Tokenization

If a model is the core component of an AI system, then data is the substrate from which it learns. The quality, quantity, and meticulous preparation of data are arguably more critical to the success of an LLM project than the specific model architecture. In this chapter, we will walk through the essential data preparation pipeline. We will cover the principles of sourcing and curating high-quality datasets, the necessary steps for cleaning and formatting that data, and the critical process of tokenization, which converts raw text into a numerical format that models can process. The entire process can be broadly classified into three phases: dataset curation and preprocessing, tokenization, and data handling with Hugging Face datasets. The data preparation workflow is illustrated in Figure 4-1.

B. K. Bolla et al., *Large Language Model Recipes*, https://doi.org/10.1007/979-8-8688-2607-8_4

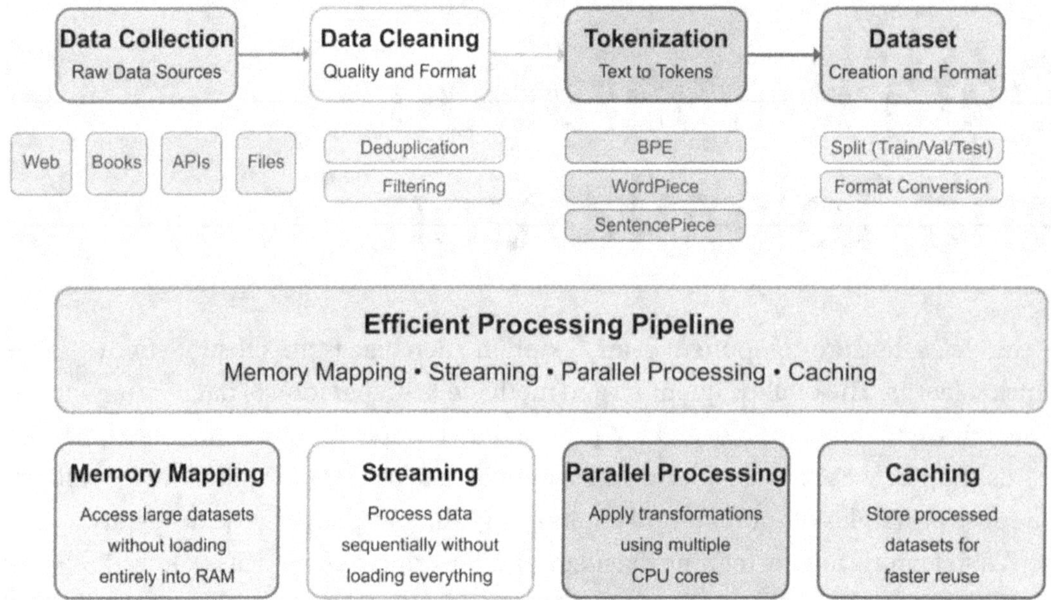

Figure 4-1. *LLM Data Preparation Workflow. This diagram illustrates the end-to-end data preparation pipeline for large language models. It shows four key stages: Data Collection (gathering raw text from diverse sources), Data Cleaning (deduplication and filtering for quality), Tokenization (converting text to tokens using algorithms like BPE), and Dataset Creation (organizing data for training). The bottom section highlights efficient processing techniques: Memory Mapping (accessing large datasets without loading entirely into RAM), Streaming (processing data sequentially), Parallel Processing (using multiple CPU cores), and Caching (storing processed data)*

Phase 1: Dataset Curation and Preprocessing

The first principle of data preparation echoes a fundamental rule: **Garbage In, Garbage Out (GIGO)**. No amount of sophisticated modeling can salvage a system built on flawed or irrelevant data. Therefore, time invested in curating high-quality data is always repaid in improved model performance. Sourcing this data involves exploring various avenues. **Public Repositories**, particularly the vast collection on the Hugging Face Hub, offer readily available datasets for numerous tasks, alongside other sources like Kaggle Datasets or academic archives. Alternatively, **Web Scraping** can be employed to gather data from online sources, although this requires careful attention to ethical considerations, robots.txt directives, and terms of service. Often, the most valuable data for specific applications comes from **Proprietary Data** sources like internal company documents or customer interactions, which demand rigorous handling regarding privacy and security.

Raw data, regardless of source, is frequently messy and requires careful cleaning. This essential step involves several procedures. **Deduplication** removes exact or near-duplicate entries, preventing models from overfitting to redundant information. For privacy and regulatory compliance (such as GDPR), **PII removal or masking** is non-negotiable; automated tools like **Microsoft Presidio**, **spaCy NER pipelines**, or **AWS Comprehend** can detect and redact names, emails, phone numbers, and addresses, though false positives and missed entities are common in proprietary datasets, making selective manual review important. **Noise reduction** strips irrelevant content such as HTML tags, boilerplate text (e.g., "Click here to unsubscribe"), or formatting artifacts, often using regular expressions or HTML parsers. Further refinement comes through **filtering**, where low-quality or irrelevant documents (very short texts, spam, or corrupted records) are removed, while **consistency checks** enforce uniform formats (such as dates and currencies) and handle missing values systematically.

Finally, the cleaned data must be structured correctly for the model's consumption, especially during fine-tuning. **Standard Formats** like JSON Lines (`.jsonl`), where each line is a JSON object representing a data sample (e.g., `{"text": "..."}`), are common and efficient. For instruction tuning, data often needs structured fields such as *instruction, input,* and *output,* following conventions like the Alpaca format. Equally critical is dividing the dataset into **train, validation, and test splits**. The training set (typically 80–90%) is used to update model weights; the validation set (5–10%) supports hyperparameter tuning and early stopping; and the test set (5–10%) is held out for final, unbiased evaluation. Before splitting, data should be **randomly shuffled** to avoid ordering bias. For **classification tasks**, stratified splitting helps preserve label distributions across splits. For **temporal or sequential datasets**, time-based splitting is essential to prevent training on future information. Finally, careful **leakage prevention**—ensuring no overlapping or highly similar samples appear across splits is crucial for reliable evaluation. Remember to shuffle your data randomly before creating these splits.

The following procedure demonstrates how to load locally stored, cleaned, and formatted data into a Dataset object, preparing it for the next stages of the pipeline.

Recipe 1: Loading Custom Ingredients (Local Files)

Goal: Load data from local CSV and JSON Lines files into a Hugging Face Dataset object.

What This Code Does (Recipe 1)

- Creates sample local CSV and JSON Lines files to simulate custom datasets

- Loads the CSV data into a Hugging Face Dataset using the load_ dataset API

- Loads a JSON Lines (.jsonl) file commonly used for instruction-tuning datasets

- Automatically assigns a default train split when loading single local files

- Displays sample records and inferred schema to verify correct loading

- Demonstrates a reusable pattern for working with local structured data in LLM pipelines

```
# --- Recipe: Loading Custom Ingredients (Local Files) ---
# Goal: Load data from local CSV and JSON Lines files into a Hugging Face
Dataset object.
# Library: Hugging Face Datasets
# Note: Ensure you have run `pip install datasets pandas`

import os
from datasets import load_dataset
import json
import csv

# --- 1. Create Dummy Data Files ---
# Create a temporary directory for our data
data_dir = "./temp_custom_data"
os.makedirs(data_dir, exist_ok=True)

# Create a dummy CSV file
csv_file_path = os.path.join(data_dir, "recipes.csv")
```

```python
csv_data = [
    {"id": 1, "dish_name": "Spaghetti Carbonara", "prep_time_mins": 15,
    "ingredients": "Pasta, Eggs, Pancetta, Cheese"},
    {"id": 2, "dish_name": "Chicken Curry", "prep_time_mins": 20,
    "ingredients": "Chicken, Onion, Tomato, Spices"},
    {"id": 3, "dish_name": "Vegetable Stir-fry", "prep_time_mins": 10,
    "ingredients": "Broccoli, Peppers, Soy Sauce, Noodles"}
]
print(f"Creating dummy CSV file: {csv_file_path}")
with open(csv_file_path, 'w', newline='', encoding='utf-8') as f:
    writer = csv.DictWriter(f, fieldnames=csv_data[0].keys())
    writer.writeheader()
    writer.writerows(csv_data)

# Create a dummy JSON Lines file (.jsonl)
jsonl_file_path = os.path.join(data_dir, "instructions.jsonl")
jsonl_data = [
    {"instruction": "Make a cup of tea.", "input": "", "output": "Boil
    water, add tea bag to cup, pour water, steep, remove bag."},
    {"instruction": "Explain photosynthesis.", "input": "", "output":
    "Photosynthesis is the process plants use to convert light energy into
    chemical energy."},
    {"instruction": "Translate to French.", "input": "Hello world",
    "output": "Bonjour le monde"}
]
print(f"Creating dummy JSON Lines file: {jsonl_file_path}")
with open(jsonl_file_path, 'w', encoding='utf-8') as f:
    for line in jsonl_data:
        f.write(json.dumps(line) + '\n')

# --- 2. Load Data using load_dataset ---

print("\n--- Loading CSV Data ---")
try:
    # For CSV, specify 'csv' type and the path to the file(s)
    # `data_files` can be a single path, a list of paths, or a dict for
        splits {'train': 'path1', 'test': 'path2'}
```

```python
    csv_dataset = load_dataset('csv', data_files=csv_file_path)
    print("CSV dataset loaded successfully:")
    print(csv_dataset)
    # Note: The default split name is 'train' when loading single files
    print("\nAccessing the 'train' split:")
    print(csv_dataset['train'])
    print("\nFirst example from CSV dataset:")
    print(csv_dataset['train'][0])
    print("\nFeatures (columns and types):")
    print(csv_dataset['train'].features)

except Exception as e:
    print(f"Error loading CSV dataset: {e}")

print("\n--- Loading JSON Lines Data ---")
try:
    # For JSON Lines, specify 'json' type
    jsonl_dataset = load_dataset('json', data_files=jsonl_file_path)
    print("JSON Lines dataset loaded successfully:")
    print(jsonl_dataset)
    print("\nFirst example from JSON Lines dataset:")
    print(jsonl_dataset['train'][0])
    print("\nFeatures (columns and types):")
    print(jsonl_dataset['train'].features)

except Exception as e:
    print(f"Error loading JSON Lines dataset: {e}")
```

Now that we have a strategy for curating and cleaning our data, the next step is to convert it into a numerical format that models can process.

Phase 2: The Tokenization Process

LLMs, being mathematical constructs, operate on numbers, not raw text. Tokenization is therefore the indispensable process of converting sequences of text into sequences of numerical IDs. It also encompasses preparing these sequences (e.g., adding special markers, padding, or truncating) to precisely match the model's input requirements.

While early NLP methods experimented with word-level or character-level tokenization, these approaches had significant drawbacks—word-level methods suffered from huge vocabularies and couldn't handle novel words (the "out-of-vocabulary" problem), while character-level methods resulted in very long sequences that obscured word semantics.

The solution adopted by virtually all modern LLMs is **Subword Tokenization**. This approach strikes an effective balance by breaking text not necessarily into words or characters but into frequently occurring sub-units. This allows the model to manage a reasonably sized vocabulary while still being able to represent rare or unknown words as sequences of known subwords (e.g., "tokenization" might become "token", "ization"). Several algorithms implement this idea. **BPE (Byte-Pair Encoding)**, used by models like GPT-2/3, starts with individual bytes and iteratively merges the most frequent adjacent pairs. **WordPiece**, favored by BERT, uses a likelihood-based criterion for merging pairs. **SentencePiece**, employed by models like Llama, Mistral, and Gemma, treats text as a stream of Unicode characters and often includes whitespace representation within tokens, making it language agnostic. The tokenization methods are compared in Figure 4-2 with an example.

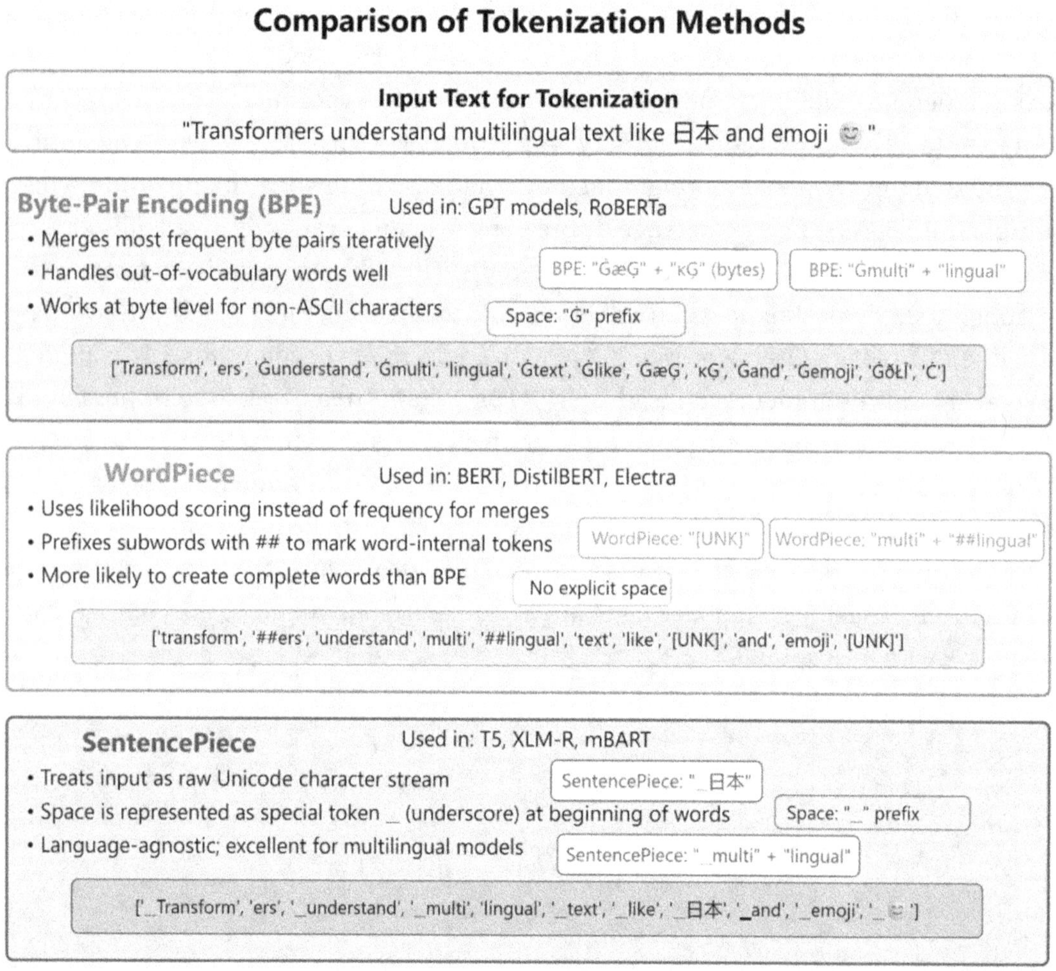

Comparison of Tokenization Methods

Figure 4-2. *Comparison of Tokenization Methods. This diagram compares how BPE, WordPiece, and SentencePiece handle the same input string, highlighting their distinct approaches to subword segmentation and special characters. BPE (used in GPT models) represents spaces with 'Ġ' and breaks non-ASCII into byte-level tokens; WordPiece (used in BERT) marks subwords with '##' and uses [UNK] for non-ASCII; SentencePiece (used in T5) marks word boundaries with '_' and handles non-ASCII directly. The comparison highlights key differences in word boundary representation, subword segmentation, and non-ASCII character handling that impact model performance across languages and text types*

Key Tokenization Concepts

Understanding **Key Tokenization Concepts** is vital for data preparation. The
Vocabulary defines the set of all possible subword tokens the tokenizer knows. Each
token is mapped to a unique integer **Token ID**. **Special Tokens** serve specific functions:
[PAD] for padding sequences to equal length in a batch, [UNK] for unknown tokens (less
common with subwords), [CLS] or [SEP] used by models like BERT, start/end sequence
markers like <s>/</s> or <|endoftext|>, and model-specific instruction tokens like
[INST]. **Padding** involves adding [PAD] tokens to shorter sequences, while **Truncation**
cuts sequences exceeding the model's maximum length. Finally, the **Attention Mask**,
a binary sequence accompanying the token IDs, is essential; it tells the model which
tokens are real (value 1) and which are padding (value 0) and should be ignored during
computation. In Figure 4-3, Byte-Pair-Encoding Tokenization-to-ID Conversion is shown
with an example in a step-by-step process.

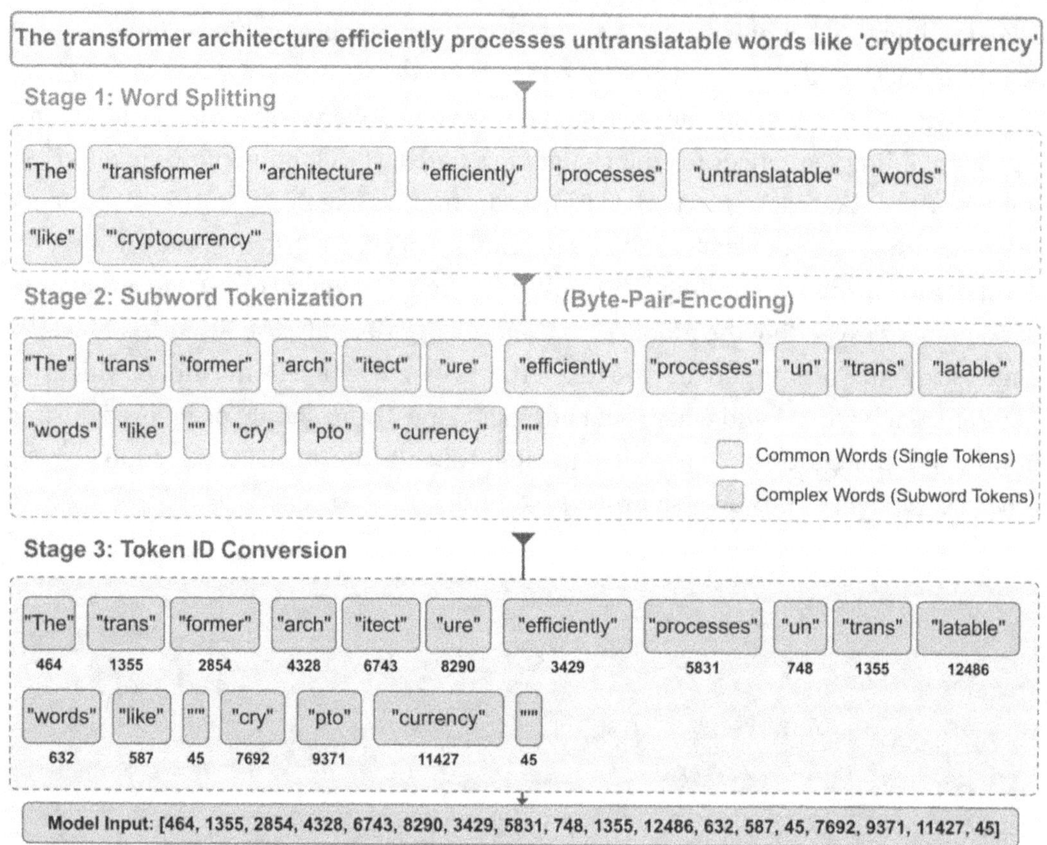

Figure 4-3. *Byte-Pair-Encoding Tokenization-to-ID Conversion. This diagram illustrates the three-stage tokenization pipeline: (1) Word Splitting separates text by whitespace and punctuation; (2) Subword Tokenization keeps common words intact (blue) while breaking complex words into meaningful subunits (orange)— e.g., "transformer" ➤ "trans" + "former"; (3) Token ID Conversion maps each token to a unique vocabulary ID. The visualization shows how tokenizers handle out-of-vocabulary words through subword segmentation and transform human-readable text into the numerical sequences processed by language models*

The following procedure provides a hands-on demonstration of these core tokenization concepts.

Recipe 2: The Tokenizer's Toolkit

Encoding, decoding, padding, truncation, and attention masks

Goal: Demonstrate core tokenizer functionalities: encoding, decoding, special tokens, padding, and truncation.

What This Code Does (Recipe 2)

- Authenticates with the Hugging Face Hub and loads a tokenizer for a chosen open source model

- Inspects tokenizer internals such as vocabulary size and special tokens (PAD, UNK, BOS, and EOS)

- Demonstrates basic **encoding** (text ➤ token IDs) and **decoding** (token IDs ➤ text)

- Shows recommended batch tokenization using padding and truncation

- Explains how attention masks distinguish real tokens from padding

- Illustrates truncation behavior when inputs exceed a defined maximum length

```
# --- Recipe: The Tokenizer's Toolkit —
# Goal: Demonstrate core tokenizer functionalities: encoding, decoding,
special tokens, padding, truncation.
# Library: Hugging Face Transformers
# Note: Ensure you have run `pip install transformers torch sentencepiece`
(SentencePiece needed for Gemma/Mistral).
# Set HF Token as ENV
import os
os.environ["HF_TOKEN"] = "hf_xxxxxxxxxxxxxxxxxxxxxx"  #replace with
your token

#Use the ENV variable
from transformers import AutoTokenizer
from huggingface_hub import HfApi, login
hf_token = os.getenv("HF_TOKEN")
```

```python
if not hf_token:
    raise ValueError("HF_TOKEN environment variable not set")
# Optional: login once (recommended)
login(token=hf_token)
# Verify identity
api = HfApi()
whoami = api.whoami(token=hf_token)
print(whoami)

# --- Configuration ---
# Choose a model ID whose tokenizer you want to explore.
# Using Gemma here as SentencePiece tokenizers are common.
model_id = "google/gemma-2b-it"
# Other options: "mistralai/Mistral-7B-v0.1", "bert-base-uncased"
(WordPiece), "gpt2" (BPE)

print(f"Loading tokenizer for model: {model_id}")
try:
    tokenizer = AutoTokenizer.from_pretrained(model_id)
    print("Tokenizer loaded successfully.")
    print(f"Tokenizer class: {tokenizer.__class__}")
    print(f"Vocabulary size: {tokenizer.vocab_size}")
    print(f"Special tokens: {tokenizer.special_tokens_map}")
    print(f"PAD token: '{tokenizer.pad_token}', ID: {tokenizer.pad_
    token_id}")
    print(f"UNK token: '{tokenizer.unk_token}', ID: {tokenizer.unk_
    token_id}")
    print(f"BOS token: '{tokenizer.bos_token}', ID: {tokenizer.bos_token_
    id}") # Beginning of Sequence
    print(f"EOS token: '{tokenizer.eos_token}', ID: {tokenizer.eos_token_
    id}") # End of Sequence
except Exception as e:
    print(f"Error loading tokenizer: {e}")
    exit()
```

```python
# --- Basic Encoding & Decoding ---
text = "LLMs need careful data preparation."
print(f"\nOriginal text: '{text}'")

# Encode: Text -> Token IDs
# `encode` is a simple method, often `tokenizer()` or `tokenizer.encode_
plus()` offer more options.
encoded_ids = tokenizer.encode(text)
print(f"Encoded IDs: {encoded_ids}")

# Decode: Token IDs -> Text
decoded_text = tokenizer.decode(encoded_ids)
print(f"Decoded text: '{decoded_text}'")

# --- Using the Tokenizer Call Method (Recommended) ---
# The __call__ method handles multiple inputs, padding, truncation, and
returns attention masks.
texts = [
    "This is a short sequence.",
    "This sequence is considerably longer and will likely exceed the
    truncation length.",
    "A third sequence."
]
print(f"\nBatch of texts:\n{texts}")

# Tokenize the batch with padding and truncation
# - padding='longest': Pad sequences to the length of the longest sequence
in the batch.
# - truncation=True: Truncate sequences longer than the model's max length
(or specified max_length).
# - max_length: Explicitly set a maximum length (optional, defaults often
to model's capability).
# - return_tensors='pt': Return results as PyTorch tensors (use 'tf' for
TensorFlow, 'np' for NumPy).
tokenized_output = tokenizer(
    texts,
    padding='longest', # Pad to longest sequence in batch
    truncation=True,   # Truncate sequences longer than max_length
```

```python
    max_length=15,       # Set an example max length
    return_tensors='pt' # Return PyTorch tensors
)

print("\nTokenized Output (Batch):")
print(f"- Input IDs:\n{tokenized_output['input_ids']}")
print(f"- Attention Mask:\n{tokenized_output['attention_mask']}") # 1 for
real tokens, 0 for PAD tokens

# --- Inspecting Padding and Attention Mask ---
print("\nInspecting the first sequence's padding:")
first_seq_ids = tokenized_output['input_ids'][0]
first_seq_mask = tokenized_output['attention_mask'][0]
print(f"- Padded IDs: {first_seq_ids.tolist()}") # .tolist() for cleaner
printing
print(f"- Corresponding Mask: {first_seq_mask.tolist()}")
# Decode manually to see the padding token (if PAD token ID is known)
# Note: decode skips special tokens by default unless skip_special_
tokens=False
print(f"- Decoded (incl. PAD if visible): '{tokenizer.decode(first_seq_ids,
skip_special_tokens=False)}'")
print(f"- Decoded (default, skips PAD): '{tokenizer.decode(first_seq_ids,
skip_special_tokens=True)}'")

# --- Inspecting Truncation ---
print("\nInspecting the second sequence's truncation:")
second_seq_ids = tokenized_output['input_ids'][1]
print(f"- Truncated IDs (length {len(second_seq_ids)}): {second_seq_ids.
tolist()}")
print(f"- Decoded truncated text: '{tokenizer.decode(second_seq_ids)}'")
# Compare length to max_length defined above (15)

# --- End of Recipe ---
```

A practical application of tokenization is analyzing the distribution of sequence lengths in a dataset. This analysis is crucial for making informed decisions about the max_length parameter during training and inference, balancing computational efficiency with information retention.

Recipe 3: Measuring Your Ingredients

Visualizing token length distributions

Goal: Tokenize a dataset sample and plot a histogram of sequence lengths to inform `max_length` decisions.

What This Code Does (Recipe 3)

- Loads a sample of a real-world text dataset for analysis

- Applies a chosen tokenizer to measure tokenized sequence lengths

- Collects token length statistics across the dataset subset

- Visualizes the distribution using a histogram for intuitive understanding

- Computes key statistics such as mean, median, and 95th percentile

- Helps select an optimal `max_length` for padding and truncation during training

```python
# --- Recipe: Measuring Your Ingredients (Visualizing Token Lengths) ---
# Goal: Tokenize a dataset sample and plot a histogram of sequence lengths.
# Libraries: Hugging Face Datasets, Transformers, Matplotlib
# Note: Ensure you have run `pip install datasets transformers torch
matplotlib sentencepiece`

import matplotlib.pyplot as plt
from datasets import load_dataset
from transformers import AutoTokenizer
import numpy as np

# --- Configuration ---
dataset_name = "imdb" # Standard sentiment analysis dataset (text reviews)
split = "train" # Choose split (e.g., 'train', 'test')
text_column = "text" # Column containing the text data in the dataset
# Choose a tokenizer consistent with models you might use
tokenizer_id = "distilbert-base-uncased" # Smaller BERT-like model
tokenizer
num_samples_for_analysis = 1000 # Analyze a subset for efficiency
```

```
print(f"Loading dataset: {dataset_name} (split: {split})")
try:
    # Load a subset using slicing for efficiency if dataset is large
    # dataset = load_dataset(dataset_name, split=f"{split}[:{num_samples_
      for_analysis}]")
    # Or load normally and select later if small enough
    full_dataset = load_dataset(dataset_name)
    dataset = full_dataset[split].select(range(min(num_samples_for_
    analysis, len(full_dataset[split])))))
    print(f"Loaded {len(dataset)} samples for analysis.")
except Exception as e:
    print(f"Error loading dataset: {e}")
    exit()

print(f"\nLoading tokenizer: {tokenizer_id}")
try:
    tokenizer = AutoTokenizer.from_pretrained(tokenizer_id)
    print("Tokenizer loaded.")
except Exception as e:
    print(f"Error loading tokenizer: {e}")
    exit()

# --- Tokenize the Dataset ---
# Define a function to tokenize examples and get lengths
def tokenize_and_get_length(batch):
    tokenized = tokenizer(batch[text_column], truncation=False,
    padding=False, max_length=512)
    lengths = [len(input_ids) for input_ids in tokenized["input_ids"]]
    return {"token_length": lengths}

print("\nTokenizing dataset samples to calculate lengths...")
try:
    # Use .map() to apply the function efficiently
    # batched=True processes multiple examples at once
    dataset_with_lengths = dataset.map(tokenize_and_get_length,
    batched=True)
    print("Tokenization complete.")
```

```python
    print(f"First few examples with token lengths:\n {dataset_with_
lengths[:3]}")
except Exception as e:
    print(f"Error during tokenization map: {e}")
    exit()

# --- Prepare Data for Plotting ---
token_lengths = dataset_with_lengths["token_length"]

# --- Plot Histogram ---
print("\nGenerating histogram of token lengths...")
plt.figure(figsize=(10, 6))
plt.hist(token_lengths, bins=50, color='skyblue', edgecolor='black') #
Adjust bins as needed
plt.title(f'Distribution of Token Lengths ({tokenizer_id} on {dataset_name}
[:{len(dataset)}])')
plt.xlabel('Token Sequence Length')
plt.ylabel('Number of Samples')
plt.grid(axis='y', alpha=0.75)

# Add some descriptive statistics
mean_len = np.mean(token_lengths)
median_len = np.median(token_lengths)
max_len = np.max(token_lengths)
percentile_95 = np.percentile(token_lengths, 95)
plt.axvline(mean_len, color='red', linestyle='dashed', linewidth=1,
label=f'Mean: {mean_len:.0f}')
plt.axvline(median_len, color='green', linestyle='dashed', linewidth=1,
label=f'Median: {median_len:.0f}')
plt.axvline(percentile_95, color='purple', linestyle='dashed', linewidth=1,
label=f'95th percentile: {percentile_95:.0f}')
plt.legend()

print(f"\nStatistics:")
print(f"- Mean length: {mean_len:.2f}")
print(f"- Median length: {median_len:.2f}")
print(f"- Max length: {max_len}")
```

```
print(f"- 95th percentile length: {percentile_95:.2f}")
print("\nWhy is this important? Helps choose appropriate 'max_length' for
truncation/padding during training.")
print("Choosing a value around the 95th percentile often balances capturing
most data vs. computational cost.")

# Show the plot
plt.tight_layout()
plt.show() # This will display the plot in a new window if run locally

# --- End of Recipe ---
```

With a comprehensive understanding of how text is tokenized and prepared for model input, we must now address the challenge of handling these datasets efficiently, especially when they are too large to fit into memory.

Phase 3: Efficient Data Handling with Hugging Face Datasets

The sheer size of modern datasets presents a significant challenge, often reaching gigabytes or terabytes—far too large to fit into a typical computer's RAM. The Hugging Face datasets library provides an elegant and efficient solution. Its efficiency stems largely from its internal use of the **Apache Arrow** format, a standardized columnar memory format optimized for analytical operations and enabling zero-copy data access between different systems.

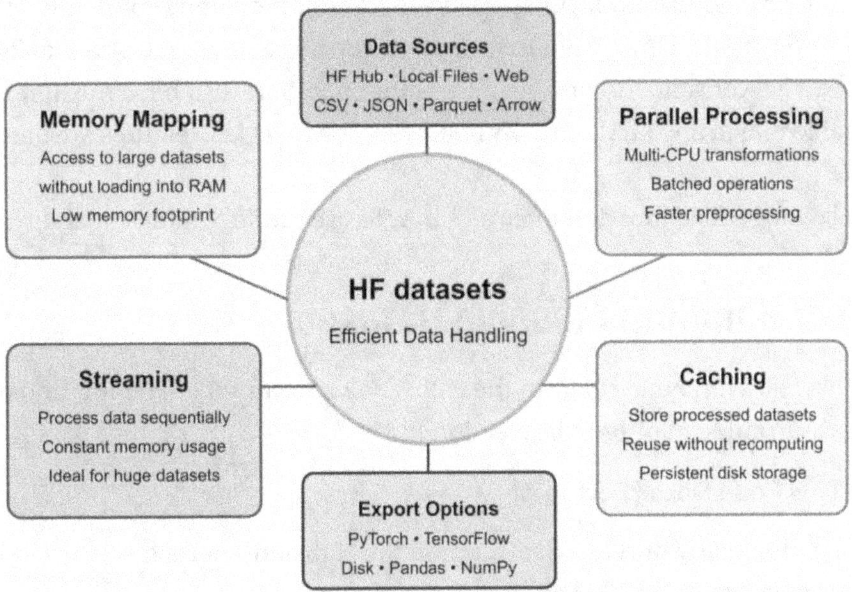

Figure 4-4. *Hugging Face Datasets Library Features. Hugging Face datasets library provides efficient data handling through multiple technical capabilities: supports diverse data sources (Hub/Local/Web) in various formats (CSV/JSON/ Parquet/Arrow), implements memory mapping for large datasets without RAM overhead, enables parallel processing with multi-CPU transformations and batched operations, offers streaming functionality for sequential processing of huge datasets with constant memory usage, includes automatic caching to store processed datasets for reuse, and provides seamless export to ML frameworks (PyTorch/TensorFlow) and data structures (Pandas/NumPy)*

This library offers several **Key Features and Benefits** crucial for handling large-scale data. By default, it employs **Memory Mapping**, accessing Arrow files on disk as if they were in RAM without needing to load the entire file upfront; the operating system handles loading necessary chunks on demand. For truly massive datasets, **Streaming** (`streaming=True`) allows iteration sample by sample directly from the source (like the Hub or local files) without downloading the complete dataset first, making it ideal for large-scale training where disk space or initial download time is a constraint. Furthermore, **Intelligent Caching** automatically saves the results of processing steps (like tokenization applied via `.map()`) to disk; subsequent runs load the cached result instantly, saving significant computation time. The library boasts a **Versatile API** including functions like `load_dataset()` for various sources, `.map()` for applying transformations efficiently (with multiprocessing support), `.filter()` for selection, `.shuffle()` for randomization, `.train_test_split()` for creating splits, and `.set_format()` for on-the-fly conversion to formats needed by ML frameworks like PyTorch or TensorFlow. This versatile API and seamless interoperability establish the library as an indispensable tool in any LLM development workflow. The key features are illustrated in Figure 4-4.

The following procedure demonstrates how to use the streaming capability.

Recipe 4: Handling Massive Datasets

Goal: Load a real-world dataset from the Hub, analyze a subset efficiently, and measure token lengths to guide max_length selection.

What This Code Does (Recipe 4)

- Loads a real-world text dataset from the Hub and selects a manageable subset for analysis

- Applies a tokenizer to each example without padding or truncation to capture true sequence lengths

- Uses the dataset `.map()` function for efficient, batched tokenization

- Extracts token length information for every sample

- Visualizes the token length distribution using a histogram

- Computes key statistics to guide `max_length` selection for model training

```
# --- Recipe: Measuring Your Ingredients (Visualizing Token Lengths) ---
# Goal: Tokenize a dataset sample and plot a histogram of sequence lengths.
# Libraries: Hugging Face Datasets, Transformers, Matplotlib
# Note: Ensure you have run `pip install datasets transformers torch
matplotlib sentencepiece`

import matplotlib.pyplot as plt
from datasets import load_dataset
from transformers import AutoTokenizer
import numpy as np
```

```python
# --- Configuration ---
dataset_name = "imdb" # Standard sentiment analysis dataset (text reviews)
split = "train" # Choose split (e.g., 'train', 'test')
text_column = "text" # Column containing the text data in the dataset
# Choose a tokenizer consistent with models you might use
tokenizer_id = "distilbert-base-uncased" # Smaller BERT-like model
tokenizer
num_samples_for_analysis = 1000 # Analyze a subset for efficiency

print(f"Loading dataset: {dataset_name} (split: {split})")
try:
    # Load a subset using slicing for efficiency if dataset is large
    # dataset = load_dataset(dataset_name, split=f"{split}[:{num_samples_
      for_analysis}]")
    # Or load normally and select later if small enough
    full_dataset = load_dataset(dataset_name)
    dataset = full_dataset[split].select(range(min(num_samples_for_
    analysis, len(full_dataset[split]))))
    print(f"Loaded {len(dataset)} samples for analysis.")
except Exception as e:
    print(f"Error loading dataset: {e}")
    exit()

print(f"\nLoading tokenizer: {tokenizer_id}")
try:
    tokenizer = AutoTokenizer.from_pretrained(tokenizer_id)
    print("Tokenizer loaded.")
except Exception as e:
    print(f"Error loading tokenizer: {e}")
    exit()

# --- Tokenize the Dataset ---
# Define a function to tokenize examples and get lengths
def tokenize_and_get_length(batch):
    tokenized = tokenizer(batch[text_column], truncation=False,
    padding=False, max_length=512)
```

```python
    lengths = [len(input_ids) for input_ids in tokenized["input_ids"]]
    return {"token_length": lengths}

print("\nTokenizing dataset samples to calculate lengths...")
try:
# Use .map() to apply the function efficiently
# batched=True processes multiple examples per call (much faster than per-
  sample processing)
# num_proc can be added for multiprocessing on CPU-heavy workloads
# For GPU pipelines, outputs can be converted directly to tensors
dataset_with_lengths = dataset.map(
    tokenize_and_get_length,
    batched=True
    # num_proc=4  # Uncomment for multiprocessing on large datasets
)
    print("Tokenization complete.")
    print(f"First few examples with token lengths:\n {dataset_with_
    lengths[:3]}")
except Exception as e:
    print(f"Error during tokenization map: {e}")
    exit()

# --- Prepare Data for Plotting ---
token_lengths = dataset_with_lengths["token_length"]

# --- Plot Histogram ---
print("\nGenerating histogram of token lengths...")
plt.figure(figsize=(10, 6))
plt.hist(token_lengths, bins=50, color='skyblue', edgecolor='black') #
Adjust bins as needed
plt.title(f'Distribution of Token Lengths ({tokenizer_id} on {dataset_name}
[:{len(dataset)}])')
plt.xlabel('Token Sequence Length')
plt.ylabel('Number of Samples')
plt.grid(axis='y', alpha=0.75)

# Add some descriptive statistics
```

```python
mean_len = np.mean(token_lengths)
median_len = np.median(token_lengths)
max_len = np.max(token_lengths)
percentile_95 = np.percentile(token_lengths, 95)
plt.axvline(mean_len, color='red', linestyle='dashed', linewidth=1,
label=f'Mean: {mean_len:.0f}')
plt.axvline(median_len, color='green', linestyle='dashed', linewidth=1,
label=f'Median: {median_len:.0f}')
plt.axvline(percentile_95, color='purple', linestyle='dashed', linewidth=1,
label=f'95th percentile: {percentile_95:.0f}')
plt.legend()

print(f"\nStatistics:")
print(f"- Mean length: {mean_len:.2f}")
print(f"- Median length: {median_len:.2f}")
print(f"- Max length: {max_len}")
print(f"- 95th percentile length: {percentile_95:.2f}")
print("\nWhy is this important? Helps choose appropriate 'max_length' for
truncation/padding during training.")
print("Choosing a value around the 95th percentile often balances capturing
most data vs. computational cost.")

# Show the plot
plt.tight_layout()
plt.show() # This will display the plot in a new window if run locally

# --- End of Recipe ---
```

Chapter Summary

In this chapter, we explored the critical processes of preparing data for Large Language
Models, emphasizing that model performance is fundamentally constrained by data
quality. We covered dataset curation, cleaning, and formatting, highlighting essential
steps such as deduplication, PII masking, noise reduction, and structured splitting for
training and evaluation. We also discussed practical preprocessing considerations,
including handling **imbalanced datasets**, applying **text normalization** (such as

lowercasing, Unicode normalization, and optional stemming or lemmatization), and cleaning **noisy user-generated content** involving emojis, HTML artifacts, or embedded code snippets.

We examined **subword tokenization** as the industry standard for converting text into numerical representations and demonstrated hands-on techniques for encoding, padding, truncation, and attention masking. Common pitfalls were highlighted, including **mismatched tokenizer–model pairs**, **inputs exceeding model sequence limits**, and **missing or unrecognized instruction-specific tokens during fine-tuning**, all of which can silently degrade model performance. Finally, we showed how modern data pipelines rely on tools like Hugging Face Datasets to efficiently process large-scale corpora using memory mapping, streaming, caching, and parallelism. With clean, well-tokenized, and efficiently managed data in place, we are now prepared to focus on prompt engineering and model interaction in the next chapter.

With our data now cleaned, tokenized, and efficiently loaded, we are ready to begin interacting with our chosen model. The next chapter will introduce the fundamental technique of prompt engineering, where we will learn how to craft effective instructions to guide model behavior and elicit desired outputs.

## PART III

Core Techniques for Prompting and Fine-Tuning

## CHAPTER 5

Prompt Engineering

Now that we have prepared our data, we are ready to communicate with our model. A Large Language Model is an incredibly knowledgeable yet literal computational system. Unlocking its full potential requires providing instructions—prompts—that are clear, precise, and well structured. **Prompt Engineering** is the essential science of crafting these inputs to guide the LLM toward generating the specific, high-quality output we desire. Unlike fine-tuning, which modifies the model's parameters, prompt engineering focuses on optimizing the input to the model, often providing the fastest route to improving performance for particular tasks. In this chapter, we will explore the fundamental techniques used to master this art of instruction.

Foundational Prompting Techniques

The most common prompting strategies involve providing the model with either a direct instruction (zero-shot prompting) or an instruction supplemented with examples (few-shot prompting).

Zero-Shot Prompting

The most straightforward approach is **zero-shot prompting**. Here, we directly ask the model to perform a task without providing any prior examples of how to accomplish it within the prompt itself. The model must rely entirely on the patterns and knowledge learned during its pre-training. This is appealing for its simplicity, but its performance can be less reliable for complex tasks or when a specific output format is required.

Recipe 1: Zero-Shot Wonders

Crafting prompts for summarization and Q&A without examples

© Bharath Kumar Bolla, Kalpa Subbaiah and Sashi Kiran Kaata 2026
B. K. Bolla et al., *Large Language Model Recipes*, https://doi.org/10.1007/979-8-8688-2607-8_5

Goal: Craft prompts for summarization and Q&A without providing any prior examples in the prompt itself. The LLM relies solely on its pre-trained knowledge.

What This Code Does (Recipe 1)

- Reads the Hugging Face access token from an environment variable and authenticates with the Hub

- Loads an instruction-tuned open source model using the Hugging Face `pipeline` API

- Automatically selects GPU execution when available and applies efficient precision settings

- Demonstrates zero-shot prompting without any task-specific fine-tuning

- Performs two common zero-shot tasks: document summarization and question answering

- Extracts and displays only the generated answers by anchoring on prompt markers

```python
# --- Recipe: Zero-Shot Wonders ---
# Goal: Demonstrate crafting zero-shot prompts for common tasks.
# Method: Using Hugging Face pipeline for simplicity (can be adapted for
API calls).

from transformers import pipeline, set_seed
from huggingface_hub import HfApi, login
import torch
import os

# Set HF Token as ENV
os.environ["HF_TOKEN"] = "hf_xxxxxxxxxxxxxxxxxxxxxx"  #replace with
your token

#Use the ENV variable
hf_token = os.getenv("HF_TOKEN")
if not hf_token:
    raise ValueError("HF_TOKEN environment variable not set")
# Optional: login once (recommended)
```

```python
login(token=hf_token)
# Verify identity
api = HfApi()
whoami = api.whoami(token=hf_token)
print(whoami)

# --- Configuration ---
# Use a model suitable for instruction following or general tasks
# Larger models generally perform better on zero-shot tasks.
# Using Gemma instruct here. Replace with Mistral Instruct, GPT variants
via API, etc.
model_id = "google/gemma-2b-it"
# Use a smaller model if resources are limited, but zero-shot quality might
decrease
# model_id = "distilgpt2"

# Use GPU if available
device_index = 0 if torch.cuda.is_available() else -1
dtype = torch.bfloat16 if torch.cuda.is_available() and torch.cuda.is_bf16_
supported() else torch.float32

print(f"Loading pipeline for model: {model_id}")
try:
    # Using text-generation pipeline; for some models/tasks, text2text-
        generation might be used.
    generator = pipeline(
        "text-generation",
        model=model_id,
        torch_dtype=dtype,
        device=device_index
    )
    print("Pipeline loaded.")
except Exception as e:
    print(f"Error loading pipeline: {e}")
    print("Ensure sufficient GPU memory if using a large model.")
    exit()
```

```python
# Set seed for reproducibility if using sampling
set_seed(42)

# --- Task 1: Zero-Shot Summarization ---
print("\n--- Zero-Shot Summarization ---")
document_to_summarize = """
Large Language Models (LLMs) are advanced artificial intelligence systems
trained on vast amounts of text data.
They excel at understanding and generating human-like text for various
tasks, including translation, summarization, question answering, and code
generation.
Key architectures like the Transformer model, utilizing mechanisms such
as self-attention, enable LLMs to capture long-range dependencies and
contextual nuances in language.
Training these models requires significant computational resources and
carefully curated datasets.
Ethical considerations, including bias mitigation and responsible
deployment, are crucial aspects of LLM development and application.
Ongoing research focuses on improving efficiency, controllability, and
reasoning capabilities of these powerful models.
"""

# Simple zero-shot prompt
prompt_summary = f"""
Summarize the following document in one sentence:

Document:
\"\"\"
{document_to_summarize}
\"\"\"

Summary:
"""

print(f"Prompt:\n{prompt_summary}")
```

```python
try:
    # Adjust max_new_tokens based on expected summary length
    outputs_summary = generator(prompt_summary, max_new_tokens=50, do_
    sample=False) # Use do_sample=False for more deterministic summary
    print("\nGenerated Summary:")
# Note: This simple string split works for demonstration purposes.
# In production systems, use structured output parsing (e.g., LangChain
OutputParsers or Pydantic schemas)
# to reliably extract model responses.    print(outputs_summary[0]
['generated_text'].split("Summary:")[-1].strip()) # Extract text after
"Summary:"
except Exception as e:
    print(f"Error during summarization: {e}")

# --- Task 2: Zero-Shot Question Answering ---
print("\n--- Zero-Shot Question Answering ---")
context = """
The Eiffel Tower, located in Paris, France, was completed in 1889 for the
Exposition Universelle (World's Fair).
It was designed and built by Gustave Eiffel's company. Initially criticized
by some of France's leading artists and intellectuals,
it has become a global icon of French culture and one of the most
recognizable structures in the world.
The tower is 330 meters (1,083 ft) tall, about the same height as an
81-story building.
"""

question = "How tall is the Eiffel Tower?"

# Simple zero-shot prompt
prompt_qa = f"""
Context:
\"\"\"
{context}
\"\"\"

Question: {question}
Answer:
"""
```

```
print(f"Prompt:\n{prompt_qa}")

try:
    # Adjust max_new_tokens based on expected answer length
    outputs_qa = generator(prompt_qa, max_new_tokens=20, do_sample=False)
    print("\nGenerated Answer:")
    print(outputs_qa[0]['generated_text'].split("Answer:")[-1].strip())
    # Extract text after "Answer:"
except Exception as e:
    print(f"Error during Q&A: {e}")
```

Few-Shot Prompting

To provide clearer guidance to the LLMs, we use **few-shot prompting**. This technique involves including a small number of examples, or "shots," directly within the prompt. These examples demonstrate the exact input/output pattern we want the model to follow. By seeing these demonstrations, the model learns the desired pattern **in context**—a phenomenon known as In-Context Learning (ICL)—and applies it to the actual query. Few-shot prompting significantly improves performance and control over the output format for many tasks compared to zero-shot, without requiring any model retraining. The trade-offs involve increased prompt length, which consumes more input tokens, and the fact that performance can be sensitive to the quality and order of the examples provided. The difference between zero-shot and few-shot is illustrated in Figure 5-1.

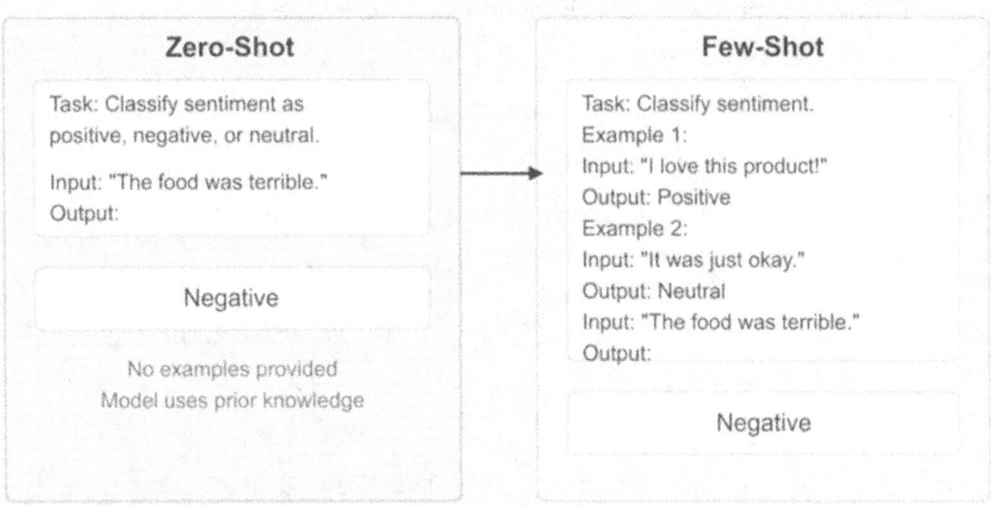

Figure 5-1. *Zero-Shot vs. Few-Shot Prompting. This diagram compares two fundamental prompting approaches: Zero-Shot (left), where the model receives only instructions without examples, and Few-Shot (right), where demonstration examples guide the model. Both approaches produce identical results for sentiment classification, but Few-Shot provides explicit pattern guidance through examples*

Recipe 2: Learning on the Fly

Implementing few-shot prompting with in-context examples

Goal: Implement few-shot prompting by including input-output examples directly within the prompt to guide the LLM's response.

Implementing few-shot prompting by including a small number (typically 1 to 5) of input-output examples directly within the prompt context to guide the LLM's response style or task execution

What This Code Does (Recipe 2)

- Creates a few-shot sentiment prompt by including three labeled examples using consistent anchors: `Input:` and `Sentiment:`

- Appends a new query and leaves `Sentiment:` blank so the model completes only the label

- Uses a small `max_new_tokens` and `do_sample=False` to encourage a single-word, stable classification

- Extracts the predicted label by splitting on the last `Sentiment:` and taking the first output line

- Reuses the same few-shot template idea for keyword extraction with `Text:` and `Keywords:` anchors

- Enables light sampling (`temperature=0.2`) for keyword extraction to allow flexible comma-separated lists

```python
# --- Task: Simple Sentiment Classification (Positive/Negative/Neutral) ---
print("\n--- Few-Shot Sentiment Classification ---")

# Define the examples (shots)
# Clear separation between input and output is key. Using "Input:" and
"Sentiment:" here.
examples = """
Input: This movie was fantastic! The acting was superb.
Sentiment: Positive

Input: The weather today is quite average, neither sunny nor rainy.
Sentiment: Neutral

Input: I'm really disappointed with the product quality. It broke after
one use.
Sentiment: Negative
"""

# Define the actual query
query_text = "The speaker delivered an engaging and informative
presentation."

# Combine examples and query into the final prompt
prompt_few_shot = f"""
Classify the sentiment of the input text as Positive, Negative, or Neutral.

{examples}
Input: {query_text}
```

```python
Sentiment:
"""

print(f"Prompt:\n{prompt_few_shot}")

try:
    # We only expect one word as output, so max_new_tokens can be small.
    # Using do_sample=False makes the output more deterministic based on
    the examples.
    outputs_few_shot = generator(prompt_few_shot, max_new_tokens=5, do_
    sample=False)
    generated_text = outputs_few_shot[0]['generated_text']

    # Extract the part after the last "Sentiment:"
    final_answer = generated_text.split("Sentiment:")[-1].strip()

    print("\nGenerated Sentiment:")
    # Sometimes models might add extra text; try to isolate the
    likely answer.
    # Split by newline and take the first line if necessary.
    print(final_answer.split('\n')[0])

except Exception as e:
    print(f"Error during few-shot generation: {e}")

# --- Task 2: Simple Format Conversion (Extracting Keywords) ---
print("\n--- Few-Shot Keyword Extraction ---")

examples_keywords = """
Text: The quick brown fox jumps over the lazy dog.
Keywords: quick, brown, fox, jumps, lazy, dog

Text: Artificial intelligence is transforming various industries.
Keywords: Artificial intelligence, transforming, industries

Text: Learn Python programming for data science and web development.
Keywords: Python, programming, data science, web development
"""
```

```python
query_keywords = "Large Language models have transformed the tech industry"

prompt_keywords = f"""
Extract the main keywords from the text, separated by commas.

{examples_keywords}
Text: {query_keywords}
Keywords:
"""

print(f"Prompt:\n{prompt_keywords}")

try:
    outputs_keywords = generator(prompt_keywords, max_new_tokens=20, do_
    sample=True, temperature=0.2) # Allow a bit of sampling
    generated_text_keywords = outputs_keywords[0]['generated_text']
    final_answer_keywords = generated_text_keywords.split("Keywords:")
    [-1].strip()
    print("\nGenerated Keywords:")
    print(final_answer_keywords.split('\n')[0])

except Exception as e:
    print(f"Error during few-shot keyword extraction: {e}")
```

Recipe 3: Head to Head

Comparing zero-shot vs. few-shot performance for a classification task

Goal: Directly compare the outputs of zero-shot and few-shot prompting for a classification task.

What This Code Does (Recipe 3)

- Defines a simple multi-class classification task using a news headline

- Constructs a zero-shot prompt with only task instructions and category options

- Builds a few-shot prompt by adding labeled example headlines

- Generates predictions using both prompting strategies under identical settings

- Extracts and compares the model's outputs for each approach

- Highlights how few-shot examples often improve accuracy and formatting

```python
# --- Recipe: Head-to-Head (Zero-Shot vs. Few-Shot) ---
# Goal: Compare the outputs of zero-shot and few-shot prompting for a
simple classification task.
# Method: Using Hugging Face pipeline.

# --- Task: Classify news headline topic ---
# Categories: Technology, Sports, Politics, Business
headline = "Stock market surges as new economic data shows strong growth."

# --- Approach 1: Zero-Shot Prompt ---
print("\n--- Zero-Shot Classification ---")
prompt_zero = f"""
Classify the topic of the following news headline. Choose from: Technology,
Sports, Politics, Business.

Headline: "{headline}"
Topic:
"""

print(f"Zero-Shot Prompt:\n{prompt_zero}")

try:
    outputs_zero = generator(prompt_zero, max_new_tokens=10, do_
    sample=False)
    answer_zero = outputs_zero[0]['generated_text'].split("Topic:")[-1].
    strip().split('\n')[0]
    print(f"\nZero-Shot Output: {answer_zero}")
except Exception as e:
    print(f"Error during Zero-Shot generation: {e}")
    answer_zero = "Error"

# --- Approach 2: Few-Shot Prompt ---
print("\n--- Few-Shot Classification ---")
few_shot_examples = """
Headline: "New iPhone model released with advanced camera features."
```

```
Topic: Technology

Headline: "Local team wins championship in thrilling overtime game."
Topic: Sports

Headline: "Parliament debates new environmental regulations bill."
Topic: Politics
"""

prompt_few = f"""
Classify the topic of the following news headline. Choose from: Technology,
Sports, Politics, Business.

{few_shot_examples}
Headline: "{headline}"
Topic:
"""
print(f"Few-Shot Prompt:\n{prompt_few}")

try:
    outputs_few = generator(prompt_few, max_new_tokens=10, do_sample=False)
    answer_few = outputs_few[0]['generated_text'].split("Topic:")[-1].
    strip().split('\n')[0]
    print(f"\nFew-Shot Output: {answer_few}")
except Exception as e:
    print(f"Error during Few-Shot generation: {e}")
    answer_few = "Error"

# --- Comparison ---
print("\n--- Comparison ---")
print(f"Headline: '{headline}'")
print(f"Zero-Shot Result: {answer_zero}")
print(f"Few-Shot Result:  {answer_few}")
print("\nObservation: Few-shot prompting often leads to more accurate or
correctly formatted results by providing clear examples.")
```

While providing examples is powerful for pattern recognition, some tasks require a more structured reasoning process. To address this, we can employ more advanced techniques that guide the model's internal thought process or even allow it to interact with external tools. The next section introduces two such methods: **Chain-of-Thought (CoT)**, which elicits step-by-step reasoning, and **ReAct (Reason + Act)**, which enables models to use tools to find information or perform calculations.

Chain-of-Thought (CoT) Prompting

For tasks that demand reasoning, such as mathematical word problems or logic puzzles, a more advanced technique called **Chain-of-Thought (CoT) Prompting** proves highly effective. CoT encourages the model to "think step by step" by explicitly generating intermediate reasoning stages before arriving at the final answer. This can be achieved in two main ways. **Few-Shot CoT** involves crafting examples where the output explicitly includes the reasoning steps. Alternatively, **Zero-Shot CoT** often works surprisingly well by simply appending a trigger phrase like "Let's think step by step" to a standard prompt. While CoT prompting increases the length of the generated output, it dramatically improves performance on complex reasoning tasks and offers the benefit of making the model's reasoning process more transparent. Chain-of-Thought of prompting is illustrated through an example in Figure 5-2.

Chain-of-Thought (CoT) Prompting

Problem: If a shirt costs $25 after a 20% discount, what was the original price?

Standard Approach

$31.25

Benefits of CoT

- Improved reasoning
- Transparency in process
- Reduced errors

Chain-of-Thought Approach

Step 1: Let's denote the original price as p.

Step 2: After a 20% discount, the price becomes 80% of the original, so 0.8 × p.

Step 3: We know that 0.8 × p = $25.

Step 4: To find p, divide both sides by 0.8: p = $25 ÷ 0.8 = $31.25.

Step 5: Therefore, the original price of the shirt was $31.25.

Problem → Reasoning Step 1 → Reasoning Step 2 → ... → Final Answer

Figure 5-2. Chain-of-Thought Prompting. This diagram contrasts the standard approach (direct answer) with the Chain-of-Thought method, which breaks down problem-solving into explicit steps. The CoT approach shows five sequential reasoning steps to calculate the original price of a discounted shirt, illustrating benefits including improved reasoning, process transparency, and reduced errors

Recipe 4: Thinking It Through

Building Chain-of-Thought prompts (zero-shot and few-shot)

Goal: Implement both zero-shot and few-shot Chain-of-Thought prompts to elicit step-by-step reasoning from the model.

Building Chain-of-Thought (CoT) prompts that explicitly encourage the LLM to "think step by step" before providing the final answer. This can be done zero-shot (e.g., adding "Let's think step by step.") or few-shot (providing examples that include reasoning steps).

What This Code Does (Recipe 4)

- Presents a reasoning-heavy math word problem to test multi-step inference

- Uses a zero-shot Chain-of-Thought prompt triggered by "Let's think step by step."

- Allows longer generation to capture intermediate reasoning and the final answer

- Introduces a few-shot Chain-of-Thought example to demonstrate the desired reasoning structure

- Applies the same reasoning pattern to a new problem via in-context learning

- Shows how few-shot CoT typically yields clearer, more reliable reasoning paths

```python
# --- Recipe: Thinking it Through (Chain-of-Thought Prompting) ---
# Goal: Demonstrate Zero-Shot and Few-Shot Chain-of-Thought prompting.
# Method: Using Hugging Face pipeline. CoT works best with larger, more
capable models.
# --- Task: Simple Math Word Problem ---
problem = """
Question: John has 5 apples. He buys 3 more boxes of apples, and each box
contains 4 apples. How many apples does John have in total?
"""


# --- Approach 1: Zero-Shot CoT ---
print("\n--- Zero-Shot Chain-of-Thought ---")

# Append the magic phrase to trigger step-by-step reasoning
prompt_zero_shot_cot = f"""
{problem}
Answer: Let's think step by step.
"""

print(f"Prompt:\n{prompt_zero_shot_cot}")
```

```python
try:
    # Allow more tokens for the reasoning steps
    outputs_zero_cot = generator(prompt_zero_shot_cot, max_new_tokens=150,
    do_sample=True, temperature=0.6)
    print("\nGenerated Response (Zero-Shot CoT):")
    # Extract the reasoning and answer part
    reasoning_answer = outputs_zero_cot[0]['generated_text'].split("Let's
    think step by step.")[-1].strip()
    print(reasoning_answer)
except Exception as e:
    print(f"Error during Zero-Shot CoT generation: {e}")

# --- Approach 2: Few-Shot CoT ---
print("\n--- Few-Shot Chain-of-Thought ---")

# Provide an example demonstrating the step-by-step reasoning format
cot_example = """
Question: A bakery made 20 cakes. They sold 15 cakes and then baked 8 more.
How many cakes do they have now?
Answer: Let's think step by step.
1. The bakery started with 20 cakes.
2. They sold 15 cakes, so they have 20 - 15 = 5 cakes left.
3. They baked 8 more cakes, so they now have 5 + 8 = 13 cakes.
Final Answer: The final answer is 13
"""

prompt_few_shot_cot = f"""
{cot_example}

{problem}
Answer: Let's think step by step.
"""

print(f"Prompt:\n{prompt_few_shot_cot}")

try:
    # Allow sufficient tokens for reasoning
```

```
    outputs_few_cot = generator(prompt_few_shot_cot, max_new_tokens=150,
    do_sample=True, temperature=0.6)
    print("\nGenerated Response (Few-Shot CoT):")
    # Extract the reasoning and answer part for the *new* problem
    reasoning_answer_few = outputs_few_cot[0]['generated_text'].
    split(problem)[-1].split("Let's think step by step.")[-1].strip()
    print(reasoning_answer_few)
except Exception as e:
    print(f"Error during Few-Shot CoT generation: {e}")
```

Agentic Frameworks: ReAct (Reason + Act)

Beyond these core methods, more advanced strategies continue to emerge. **Self-Consistency**, for instance, builds upon CoT by generating multiple reasoning chains for the same prompt and then selecting the most frequent final answer, improving robustness against errors. Another powerful concept is **ReAct (Reason + Act)**, which aims to combine the reasoning capabilities of CoT with the ability to take actions, such as using external tools like a calculator or performing a web search via an API. The model learns to interleave reasoning steps with action steps and observations, paving the way for more capable and autonomous AI agents, often facilitated by frameworks like LangChain or LlamaIndex. Self-Consistency and ReAct prompting strategies are contrastively illustrated in Figure 5-3 with an example. A detailed step-by-step approach of ReAct prompting is illustrated in Figure 5-4.

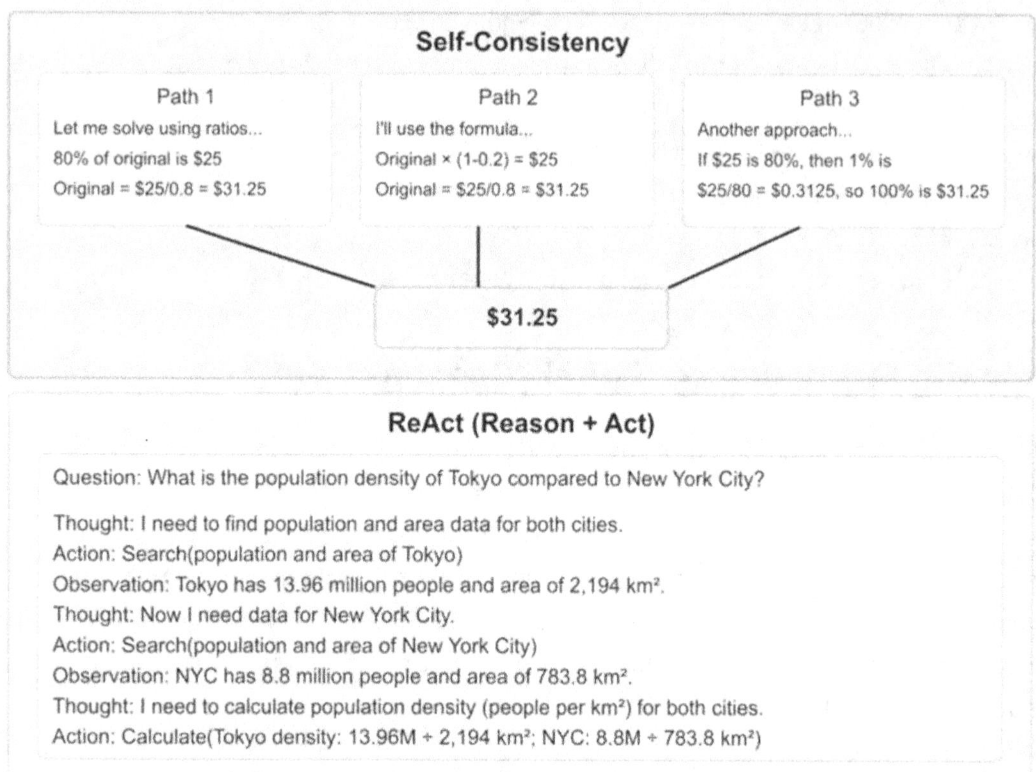

Figure 5-3. *Advanced Prompting Strategies. This diagram illustrates two key techniques: Self-Consistency, which generates multiple reasoning paths to reach a consistent answer ($31.25) (top), and ReAct (Reason + Act), which combines reasoning with tool use in a structured thought-action-observation cycle for solving complex queries requiring external information (bottom)*

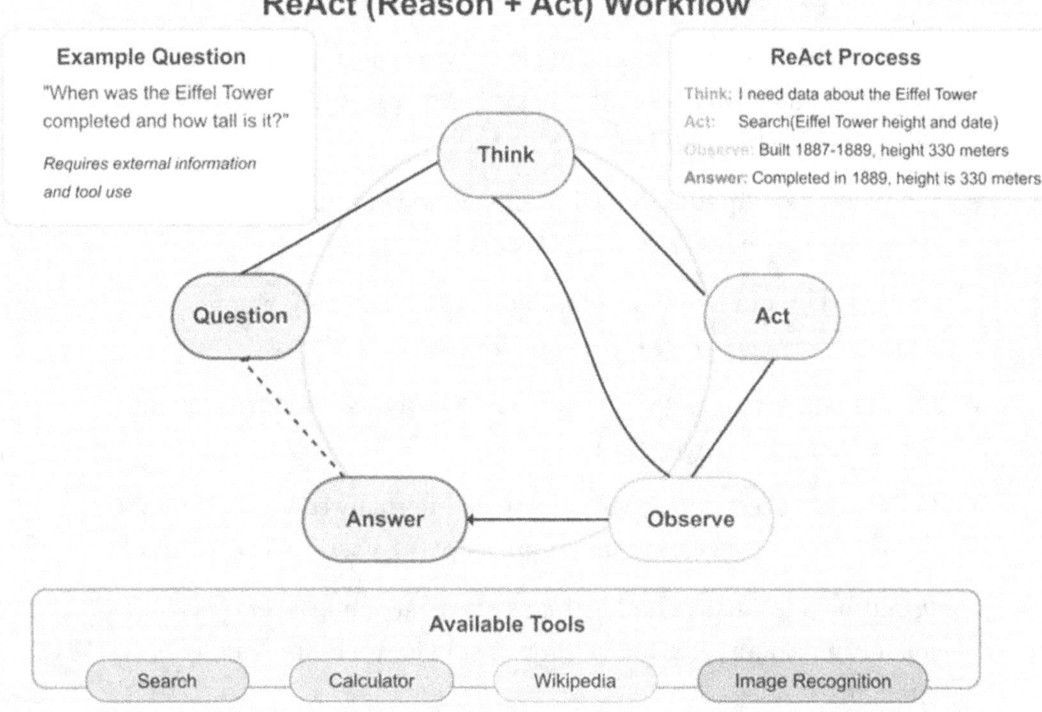

Figure 5-4. *ReAct Workflow. This diagram illustrates the ReAct (Reason + Act) process, showing how language models can combine reasoning with tool use. The cycle moves from Question to Think to Act to Observe to Answer, with a concrete example of retrieving Eiffel Tower information. Available tools (Search, Calculator, Wikipedia, and Image Recognition) enable the model to access external information needed to solve complex queries*

Recipe 5: ReActing to the World-Using Tools with LangChain Agents

Goal: Implement the ReAct pattern using the LangChain framework, allowing an LLM to use an external tool (Wikipedia) to answer a question.

Implementing the ReAct (Reason + Act) framework using LangChain agents. This recipe shows how an LLM can interact with external tools (like search engines or calculators) by iteratively generating thoughts about the problem, choosing an action (a tool to use), providing input to the action, and observing the result to inform its next step toward the final answer.

What This Code Does (Recipe 5)

- Loads an instruction-tuned model and wraps a Hugging Face generation pipeline into a LangChain-compatible LLM (HuggingFacePipeline)

- Initializes a **Wikipedia tool** with constrained output (top_k_results, doc_content_chars_max) and registers it in tools

- Loads a standard ReAct prompt from LangChain Hub (or falls back to a local prompt template if Hub pull fails)

- Creates a ReAct agent by binding **LLM + tools + ReAct prompt** and then wraps it in an AgentExecutor

- Calls agent_executor.invoke(), which internally runs the **Thought → Action → Observation loop** until a final answer (or limits/errors)

- Prints the final answer and includes practical debug levers (verbose tracing, max iterations, tool additions, and temperature/model strength)

```
pip install langchain langchain-huggingface transformers torch accelerate
sentencepiece wikipedia langchain_community

# --- Recipe: ReAct with LangChain (Using Built-in Tools) ---
# Goal: Implement the ReAct pattern using the LangChain framework with
built-in tools.
# Method: Uses LangChain agents, built-in Wikipedia tool, and an LLM
wrapper.
# Libraries: langchain, langchain-huggingface (or other LLM provider),
#            transformers, torch, accelerate, sentencepiece, wikipedia
# Note: Install required libs: pip install langchain langchain-huggingface
transformers torch accelerate sentencepiece wikipedia

import torch
from transformers import pipeline, AutoTokenizer, AutoModelForCausalLM
from langchain_huggingface import HuggingFacePipeline # LLM Wrapper
from langchain.agents import AgentExecutor, create_react_agent, Tool
# Agent components
```

```python
from langchain_core.prompts import PromptTemplate # For ReAct prompt
internal to LangChain
# --- Import Built-in Tool ---
from langchain_community.tools import WikipediaQueryRun
from langchain_community.utilities import WikipediaAPIWrapper
# --- End Import ---
import re
import math
import os

# --- 1. Configuration & Setup ---
# Use an instruction-tuned model
MODEL_ID = "google/gemma-2b-it" # Needs good instruction following

# Use GPU if available
device_index = 0 if torch.cuda.is_available() else -1
dtype = torch.bfloat16 if torch.cuda.is_available() and torch.cuda.is_bf16_
supported() else torch.float32
MAX_ITERATIONS = 5 # Limit agent steps

# --- 2. Load LLM (via Pipeline Wrapper) ---
# LangChain needs an LLM interface. We wrap the HF pipeline.
print(f"Loading pipeline for model: {MODEL_ID}")
try:
    tokenizer_lc = AutoTokenizer.from_pretrained(MODEL_ID)
    model_lc = AutoModelForCausalLM.from_pretrained(
        MODEL_ID,
        torch_dtype=dtype,
        # device_map="auto" # Use device map for larger models if needed
    )
    model_lc.to(f'cuda:{device_index}' if device_index >= 0 else 'cpu')

    if tokenizer_lc.pad_token is None:
        tokenizer_lc.pad_token = tokenizer_lc.eos_token
    if model_lc.config.pad_token_id is None:
        model_lc.config.pad_token_id = tokenizer_lc.pad_token_id
```

```python
    # Create HF pipeline
    # Why: LangChain agents expect an LLM interface. We wrap a HF
    pipeline so the agent can call it.
    # Why max_new_tokens=250: ReAct needs room for multiple
    Thought/Action steps + a final answer.

    pipe = pipeline(
        "text-generation",
        model=model_lc,
        tokenizer=tokenizer_lc,
        max_new_tokens=250,
        temperature=0.6, # Control randomness for agent
        pad_token_id=tokenizer_lc.eos_token_id # Important for generation
    )

    # Wrap pipeline for LangChain
    llm = HuggingFacePipeline(pipeline=pipe)
    print("LangChain LLM Wrapper created.")

except Exception as e:
    print(f"Error loading model/pipeline for LangChain: {e}")
    exit()

# --- 3. Define Tools (Using Built-in Wikipedia) ---
# Instantiate the Wikipedia API Wrapper and Tool
# You can customize top_k_results, doc_content_chars_max etc.
print("\nInitializing built-in tools...")
try:
    api_wrapper = WikipediaAPIWrapper(top_k_results=1, doc_content_chars_
    max=1000) # Limit context length
    wiki_tool = WikipediaQueryRun(api_wrapper=api_wrapper)
    # The tool object automatically gets name='wikipedia' and a
      description.
    # You can override if needed, but defaults are usually fine.
    print(f"Tool Name: {wiki_tool.name}")
    print(f"Tool Description: {wiki_tool.description}")
```

```python
except ImportError:
    print("Error: 'wikipedia' library not found. Please install it: pip
    install wikipedia")
    exit()
except Exception as e:
    print(f"Error initializing Wikipedia tool: {e}")
    exit()

# Define the list of tools the agent can use
tools = [wiki_tool]

# If you needed a calculator, you could add:
# from langchain.chains.llm_math.base import LLMMathChain
# calculator_tool = Tool( name="Calculator", func=LLMMathChain.from_
llm(llm=llm).run, description="Useful for when you need to answer questions
about math." )
# tools = [wiki_tool, calculator_tool]

print(f"\nTools available to agent: {[tool.name for tool in tools]}")

# --- 4. Create ReAct Agent Prompt ---
# Pull the standard ReAct prompt template from LangChain Hub.
print("\nLoading ReAct prompt template from LangChain Hub...")
try:
    from langchain import hub
    react_prompt = hub.pull("hwchase17/react") # Pulls a standard
    ReAct prompt
    print("Loaded standard ReAct prompt template.")
except Exception as e:
    print(f"Error pulling prompt from hub: {e}. Using basic fallback.")
    # Define a basic fallback template (might be less effective than hub
      version)
    react_prompt_template_str = """Answer the following questions as best
    you can. You have access to the following tools:

{tools}
```

```
Use the following format:

Question: the input question you must answer
Thought: you should always think about what to do
Action: the action to take, should be one of [{tool_names}]
Action Input: the input to the action
Observation: the result of the action
... (this Thought/Action/Action Input/Observation can repeat N times)
Thought: I now know the final answer
Final Answer: the final answer to the original input question

Begin!

Question: {input}
Thought:{agent_scratchpad}"""
    react_prompt = PromptTemplate.from_template(react_prompt_template_str)

# --- 5. Create ReAct Agent ---
# This binds the LLM, tools, and prompt together.
print("\nCreating ReAct agent...")
try:
    agent = create_react_agent(llm, tools, react_prompt)
    print("ReAct agent created.")
except Exception as e:
    print(f"Error creating ReAct agent: {e}")
    exit()

# --- 6. Create Agent Executor ---
# The executor runs the agent loop (Thought->Action->Observation->...)
print("\nCreating Agent Executor...")
agent_executor = AgentExecutor(
    agent=agent,
    tools=tools,
    verbose=True, # Set to True to see the agent's thoughts and actions
    handle_parsing_errors=True, # Try to gracefully handle LLM output
    parsing errors
    #max_iterations=MAX_ITERATIONS
)
print("Agent Executor created.")
```

```python
# --- 7. Run the Agent ---
# --- Updated Question ---
question = "when was the Eiffel tower completed and how tall is it?"
print(f"\n--- Running Agent for Question: {question} ---")

try:
    # Use invoke for the main execution call
    # AgentExecutor.invoke() runs the full ReAct loop:
    # Thought -> Action -> Observation (repeated internally by LangChain)

    response = agent_executor.invoke({"input": question})
    # The final structured output is returned once the agent finishes
    reasoning
    final_answer = response.get("output", "Agent did not return an output.")
    print("\n==============================")
    print(f"Final Answer from Agent: {final_answer}")
    print("==============================")

except Exception as e:
    # If the agent fails, common causes include:
    # - Malformed tool output or parsing errors
    # - The agent exceeding its reasoning budget
    # - Tool failures (API issues, empty responses)
    print(f"Error during agent execution: {e}")

print("\nNote: LangChain handles the prompt formatting, parsing, and loop
execution using built-in tools.")
print("Reliability still depends heavily on the base LLM's ability.")

# If the agent produces malformed output, enable verbose=True to inspect
Thought/Action steps
# If the agent stops early or loops, increase max_iterations in
AgentExecutor.
# For better reasoning, add tools incrementally (e.g., calculator) rather
than many at once.
# If answers are unstable, lower temperature or use a stronger instruction-
tuned model.

# --- End of Recipe ---
```

While these specific strategies are powerful, they all benefit from a set of general principles that apply to almost any prompt engineering effort.

General Best Practices for Prompting

Regardless of the specific technique, effective prompt engineering benefits from several general principles:

- **Clarity and Specificity:** Be as direct and unambiguous as possible.

- **Assign a Persona:** Instructing the model to act in a specific role (e.g., "Act as an expert Python programmer") can significantly shape the tone and content of its response.

- **Specify the Output Format:** Explicitly request a particular format (e.g., "Provide the answer as a JSON object with keys 'item' and 'price'") to get structured data.

- **Use Delimiters:** Use markers like triple backticks (```), XML tags (<example>), or other clear separators to distinguish instructions from context or user input.

- **Decompose Complex Tasks:** Break down intricate tasks into simpler sub-tasks that can be prompted sequentially.

- **Iterate and Refine:** Prompt engineering is fundamentally an iterative process of testing, analyzing outputs, and refining the input.

Chapter Summary

This chapter explored the art and science of prompt engineering, the practice of crafting effective inputs to guide LLM behavior. We learned that techniques range from the simple **zero-shot** request to the more explicit **few-shot** prompting, which leverages a model's in-context learning ability. For more complex problems, **Chain-of-Thought (CoT)** prompting elicits step-by-step reasoning, and advanced strategies like **ReAct** can even enable models to use external tools.

Having mastered the art of guiding a pre-trained model through its input, we now turn our attention to a more intensive process: modifying the model's internal parameters through fine-tuning to specialize its behavior even further.

Full-Parameter Fine-Tuning

Having become adept at guiding pre-trained models with carefully crafted prompts, we now turn to a more intensive and powerful method of specialization. When a model must truly speak the language of a specific domain, understand a unique task, or adopt a particular style, fine-tuning becomes essential. In this chapter, we will delve into **Full Fine-Tuning**, the traditional and most thorough method, in which we meticulously adjust all model parameters to align the model with our desired outcome.

The Principle of Transfer Learning in LLMs

The foundational concept that makes fine-tuning possible is **transfer learning**. Building a massive language model from scratch is a monumental task, requiring large datasets and substantial computational resources. Transfer learning provides an effective shortcut by leveraging the general knowledge already encoded in large-scale pre-trained models. The concept of transfer learning is illustrated in Figure 6-1.

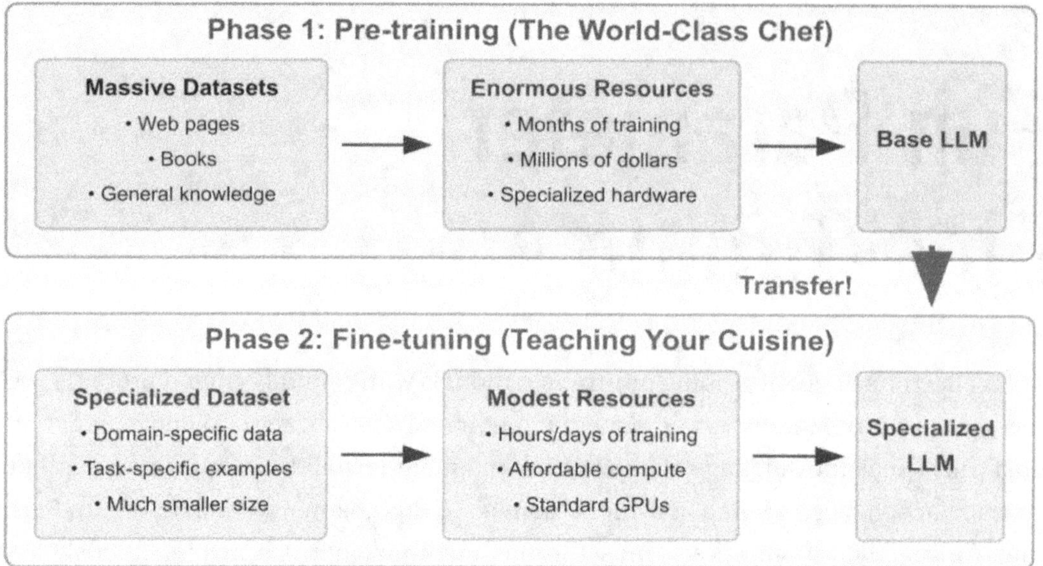

Figure 6-1. *The Two-Phase Process of Transfer Learning in LLMs. This diagram illustrates how LLM development is split into distinct phases: pre-training (top) requires massive datasets, enormous computational resources, and months of training to create a general-purpose base model. Fine-tuning (bottom) then efficiently transfers this knowledge using smaller domain-specific datasets, modest computing resources, and much shorter training times to create specialized models. This approach dramatically reduces the resources needed for tailoring LLMs to specific tasks*

The core idea is to begin with a model that already possesses a vast, general understanding of language. We then "transfer" this knowledge by continuing its training on a much smaller, specialized dataset. An LLM's initial layers grasp the fundamentals of language (grammar and syntax), while deeper layers capture more abstract concepts. Fine-tuning primarily adapts these deeper layers to the nuances of the new task without forcing the model to relearn the basics. This results in a dramatic reduction in the required data and computational power compared to pre-training from scratch, enabling us to achieve significantly better performance on our target tasks. Figure 6-2 illustrates the concept of how different layers change after the fine-tuning process.

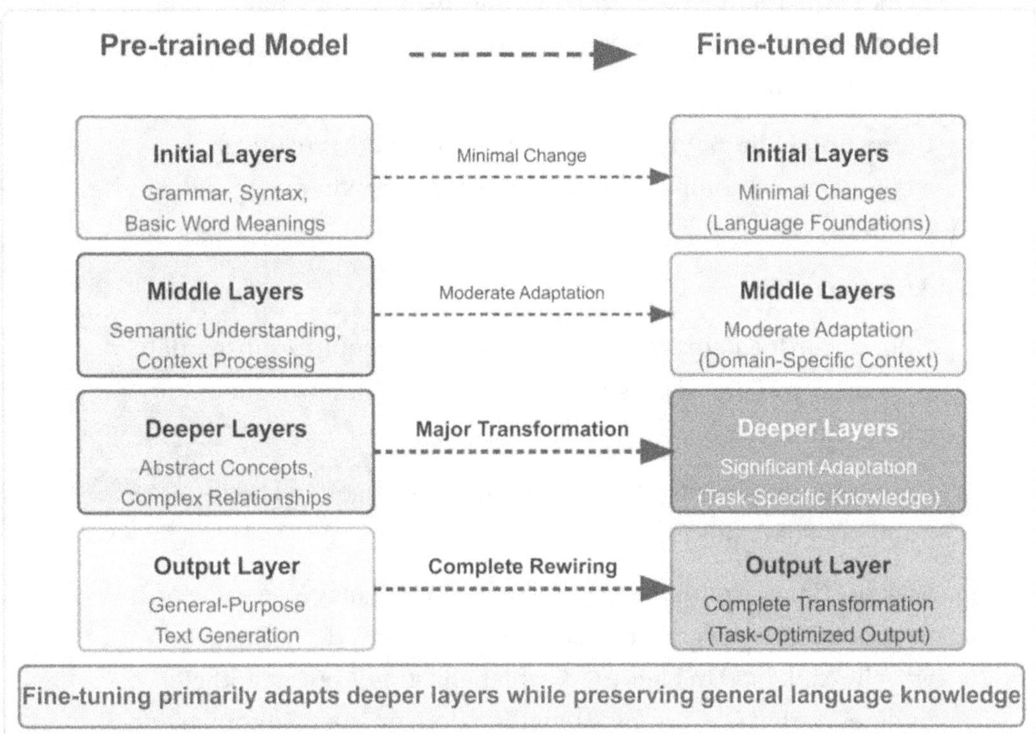

Figure 6-2. *Differential Adaptation of LLM Layers. This diagram illustrates how different layers of an LLM adapt during fine-tuning. Initial layers (handling grammar and syntax) undergo minimal changes, preserving fundamental language capabilities. Middle layers experience moderate adaptation for domain context, while deeper layers undergo major transformation to encode task-specific knowledge. The output layer is completely rewired for the target task. This layered adaptation explains why fine-tuning is efficient—it modifies what's necessary while maintaining general linguistic knowledge*

The Full Fine-Tuning Method

Full fine-tuning involves unfreezing **all** the layers and weights of the pre-trained model, allowing every single parameter to be updated during the training process on our specialized dataset.

Advantages

- **Peak Performance Potential:** By allowing all parameters to adapt, the model maximizes the likelihood of achieving the highest possible accuracy on the target task.

- **Conceptual Simplicity:** Updating all parameters is relatively straightforward compared with more complex, parameter-efficient methods.

Considerations

- **High Resource Requirements:** This method requires substantial GPU memory (VRAM) to store gradients and optimizer states for each parameter.

- **Computationally Intensive:** Training takes longer due to the large number of parameters.

- **Risk of "Catastrophic Forgetting":** The model may lose some of its general knowledge as it specializes in the new data. This risk is visually explained in Figure 6-3, which illustrates how parameter updates during fine-tuning strengthen domain-specific connections while weakening general knowledge connections. This risk can be mitigated by combining robust techniques: using small learning rates, applying regularization methods (such as Elastic Weight Consolidation or L2-SP), employing data-mixing strategies (to retain some general-domain data), and adjusting the training schedule (e.g., checkpoint averaging or gradual unfreezing).

Understanding these trade-offs is key to determining if this resource-intensive method is the correct approach for a given project.

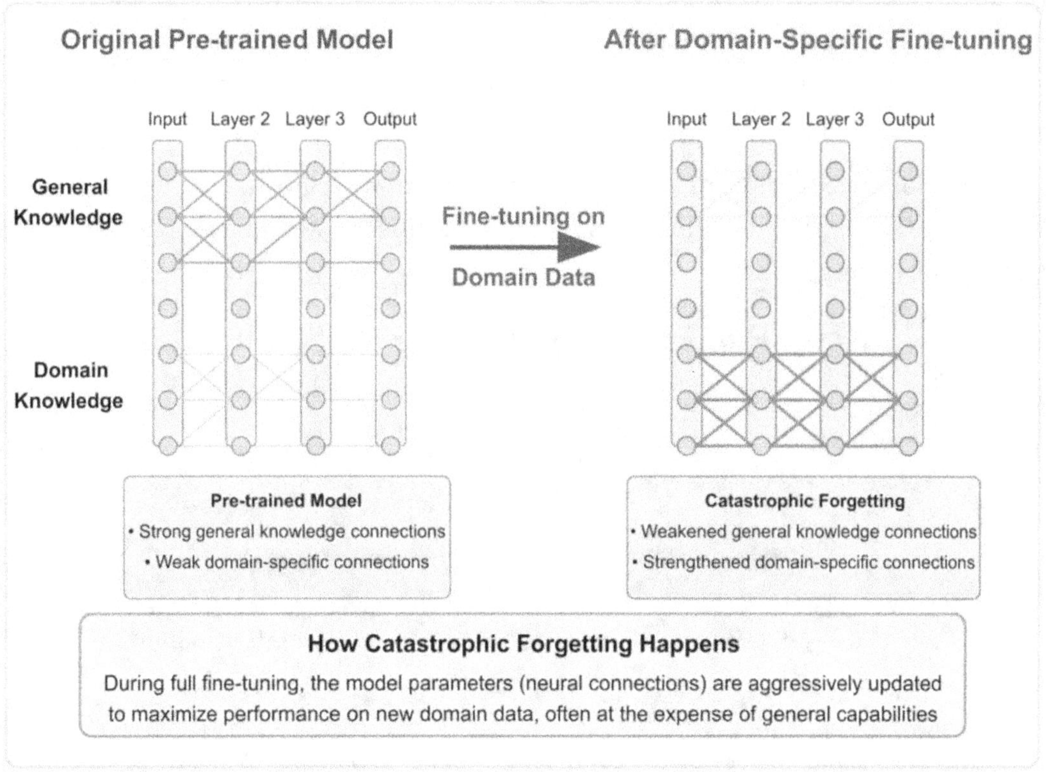

Figure 6-3. *Catastrophic Forgetting in Neural Networks. During full fine-tuning, parameter updates strengthen domain-specific connections (bottom) while weakening general knowledge connections (top). This trade-off explains why fine-tuned models may gain specialized capabilities at the expense of broader knowledge and skills*

When Is Full Fine-Tuning the Right Choice?

Consider reaching for this technique when

- Your target task or data domain is substantially different from the general text on which the model was originally trained.

- Achieving the absolute maximum performance on your specific task is the top priority, and resources are less of a constraint.

- You have access to sufficient computational power, particularly GPUs with ample VRAM.

- You want to establish a performance baseline against which to compare more resource-efficient methods, such as PEFT, discussed in subsequent chapters.

A decision tree to guide the selection of full fine-tuning is illustrated in Figure 6-4. Once we've decided that full fine-tuning is the appropriate path, the next step is to gather the necessary components.

When Full Fine-Tuning Is the Right Choice

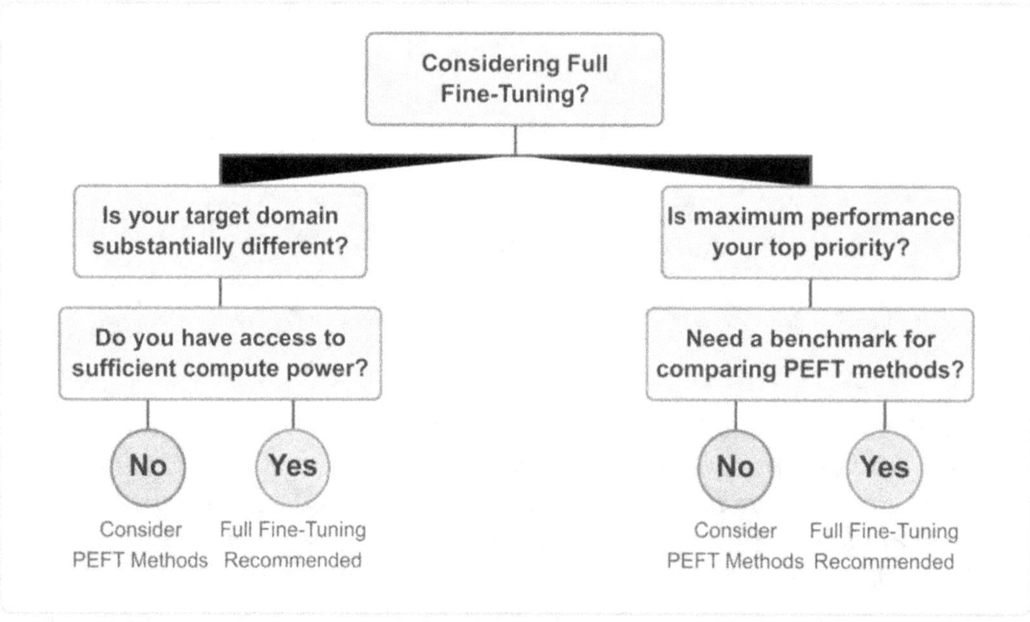

Figure 6-4. *Decision Tree for Full Fine-Tuning. This decision tree guides practitioners on when to use full fine-tuning vs. parameter-efficient methods. Full fine-tuning is recommended when you have a substantially different target domain and sufficient compute resources, or when maximum performance is the priority, or you need a benchmark for comparing PEFT methods. Otherwise, consider more efficient alternatives*

The Fine-Tuning Workflow and Hyperparameters

Successful fine-tuning requires several key components:

- **Pre-trained Model:** The foundation. Choose a base model (e.g., distilbert-base-uncased, gemma-2b) suitable for your task complexity and available resources.

- **Tokenizer:** The **exact** tokenizer that corresponds to your chosen pre-trained model. Using a mismatched tokenizer will lead to poor results.

- **Task-Specific Data:** Your specialized dataset. This needs to be high-quality, clean data, meticulously formatted for your specific task and split into distinct training, validation, and test sets.

The Fine-Tuning Workflow

The process generally follows these steps, which we'll detail in the code procedures:

1. **Load Data:** Use libraries like Hugging Face datasets to load your training, validation, and test data splits.

2. **Load Tokenizer:** Fetch the correct AutoTokenizer associated with your chosen model.

3. **Preprocess and Tokenize:** Apply the tokenizer to your text data. This involves converting text to numerical IDs, handling padding (making sequences the same length) and truncation (cutting off sequences that are too long), ensuring columns are correctly named (e.g., "labels" for classification), and converting the data into the format expected by your deep learning framework (like PyTorch tensors). A `DataCollator` often handles dynamic padding efficiently during batch creation.

4. **Load Model:** Load the pre-trained model using the appropriate Hugging Face `AutoModel` class for your task (e.g., `AutoModelForSequenceClassification`, `AutoModelForCausalLM`). This ensures the model has the correct output layer (head) for the job.

5. **Define Data Collator:** Select a suitable data collator (e.g., DataCollatorWithPadding for classification/token classification, DataCollatorForLanguageModeling for generative tasks) to efficiently handle batching.

6. **Define Evaluation Metrics:** Select metrics appropriate for evaluating your task (e.g., accuracy, F1 score for classification; ROUGE, BLEU, perplexity for generation; seqeval metrics for token classification). Use libraries like `evaluate` to create a `compute_metrics` function that the `Trainer` can use.

7. **Configure Training Arguments:** Use the `TrainingArguments` class to set key hyperparameters such as learning rate, batch size, and number of epochs and to configure logging, model checkpointing frequency, and output directories.

8. **Initialize Trainer:** Create an instance of the Hugging Face `Trainer` that aggregates the model, training arguments, datasets, tokenizer, data collator, and your `compute_metrics` function.

9. **Train:** Start the fine-tuning process by calling `trainer.train()`.

10. **Evaluate:** Assess the model's performance on the validation set (during or after training) and finally on the held-out test set using `trainer.evaluate()`.

11. **Save:** The `Trainer` typically handles saving model checkpoints during training. Ensure the final, best-performing model is saved for later use.

Finding the right settings—the **hyperparameters**—is an iterative process critical to success:

- **Learning Rate:** Arguably the most critical. Set it too high, and the training process can become unstable and diverge; set it too low, and training will be painfully slow or might get stuck in a suboptimal state. For full fine-tuning, learning rates are typically quite small, often in the range of `1e-5` to `5e-5` (0.00001 to 0.00005). Using a learning rate scheduler (such as the default linear decay in Trainer) helps adjust the learning rate during training for improved stability.

- **Batch Size:** The number of data samples the model processes in a single pass. It's primarily limited by your GPU's memory. Larger batches can provide more stable gradient estimates but consume more VRAM. If you're memory constrained, you can use `gradient_accumulation_steps` in `TrainingArguments` to simulate the effect of a larger batch size by accumulating gradients over several smaller steps before updating the model weights. Common batch sizes range from 8 to 64.

- **Number of Epochs:** The number of times the model passes over the entire training dataset. For fine-tuning, you typically need only a few epochs (typically 1 to 5) to avoid overfitting, in which the model learns the training data too well and performs poorly on unseen data. Keep a close eye on the validation loss and metrics—if they stop improving or start getting worse, it's time to stop training (the `Trainer` can be configured for early stopping).

- **Weight Decay:** A regularization technique (often L2 regularization, used by the default AdamW optimizer) that helps prevent the model's weights from becoming too large. This discourages overfitting. Common values are small, like 0.01 or 0.1.

The Golden Rule: Always tune your hyperparameters based on performance on the **validation set**. The **test set** should remain untouched until the very end for a final, unbiased evaluation.

Encoder Fine-Tuning

Encoder Fine-Tuning (as with DistilBERT) is the classic and most resource-efficient approach for tasks that require understanding the entire sentence context, such as text classification or Named Entity Recognition (NER). Their bidirectional nature makes them specialists at these "Natural Language Understanding" (NLU) tasks. They are an excellent starting point and a strong baseline.

The following procedure demonstrates this workflow by fine-tuning a classic encoder-based model, DistilBERT, for multi-class text classification.

Recipe 1: Multi-class Classification (DistilBERT)

Fine-tuning the encoder-based DistilBERT on the AG News topic classification dataset

Goal: Fully fine-tune the encoder-based DistilBERT model for multi-class text classification on the AG News dataset.

This recipe demonstrates the classic, most resource-efficient method for Full Fine-Tuning: adapting an encoder-based model, DistilBERT, to a Natural Language Understanding (NLU) task. Encoder models are specialized in capturing the full context of a sentence and are well suited for classification. This procedure establishes a strong performance baseline before considering more complex or resource-intensive architectures.

The fine-tuning workflow for this recipe follows these general steps:

1. **Load Data:** Load the AG News dataset and create a standard training, validation, and test split.Ru.

2. **Load Tokenizer and Preprocess:** Load the correct distilbert-base-uncased tokenizer and apply it to the text, handling truncation and removing the original text column.

3. **Load Model:** Load the pre-trained model using AutoModelForSequenceClassification, which automatically adds a classification head for the 4 AG News classes (World, Sports, Business, and Sci/Tech).

4. **Define Training:** Configure TrainingArguments with a small learning rate (2e–5) and define an accuracy-based compute_ metrics function.

5. **Execute and Evaluate:** Initialize the Hugging Face Trainer to run the training process, evaluating performance on the validation set after each epoch and finally on the held-out test set.

What to Expect

After training for a few epochs (typically 2–3), a fully fine-tuned DistilBERT model on the AG News dataset can achieve approximately 93–94% accuracy on the held-out test set, confirming its status as a strong baseline for text classification.

```
# --- Recipe: Multi-Class Classification (AG News / DistilBERT) ---
# Goal: Full fine-tuning of DistilBERT for multi-class text classification.
```

```python
# Dataset: AG News (4 classes: World, Sports, Business, Sci/Tech)
# Libraries: transformers, datasets, evaluate, torch
# Note: Ensure libraries are installed: pip install transformers datasets
evaluate accelerate torch

from huggingface_hub import HfApi
from huggingface_hub import login

api = HfApi()
whoami = api.whoami(token="hf_xxxxxxxxxxxxxxxxxxxxxx")
print(whoami)
login("hf_xxxxxxxxxxxxxxxxxxxxxx")

import transformers
print("Transformers version:", transformers.__version__)
print("TrainingArguments source:", transformers.TrainingArguments.__
module__)

import torch
import evaluate
import numpy as np
from datasets import load_dataset
from transformers import (
    AutoTokenizer,
    AutoModelForSequenceClassification,
    TrainingArguments,
    Trainer,
    DataCollatorWithPadding

)


# --- Configuration ---
MODEL_CHECKPOINT = "distilbert-base-uncased" # Baseline model
DATASET_NAME = "ag_news" # Multi-class news topic classification
NUM_EPOCHS = 1 # Use 1 epoch for quick demo, 2-3 might yield better results
BATCH_SIZE = 16 # Adjust based on GPU memory
LEARNING_RATE = 2e-5
OUTPUT_DIR = "./fine_tune_distilbert_agnews_output" # Directory for this model
```

```python
# --- 1. Load Data ---
print(f"Loading dataset: {DATASET_NAME}")
try:
    raw_datasets = load_dataset(DATASET_NAME)
    # AG News has 'train' and 'test' splits. Create a validation split
        from train.
    train_val_split = raw_datasets["train"].train_test_split(test_size=0.1,
        seed=42)
    raw_datasets["train"] = train_val_split["train"]
    raw_datasets["validation"] = train_val_split["test"]
    print("Dataset loaded and split:")
    print(raw_datasets)
    # Get label mapping
    label_feature = raw_datasets["train"].features["label"]
    num_labels = label_feature.num_classes
    id_to_label = {i: label for i, label in enumerate(label_feature.names)}
    print(f"Number of labels: {num_labels}")
    print(f"Label mapping: {id_to_label}")
except Exception as e:
    print(f"Error loading dataset: {e}")
    exit()

# --- 2. Load Tokenizer ---
print(f"\nLoading tokenizer for checkpoint: {MODEL_CHECKPOINT}")
try:
    tokenizer = AutoTokenizer.from_pretrained(MODEL_CHECKPOINT)
except Exception as e:
    print(f"Error loading tokenizer: {e}")
    exit()

# --- 3. Preprocess Data ---
def tokenize_function(examples):
    # Tokenize text (column name is 'text' in ag_news)
    return tokenizer(examples["text"], truncation=True, padding=False)

print("\nTokenizing dataset...")
try:
```

```python
    tokenized_datasets = raw_datasets.map(tokenize_function, batched=True)
    # Rename 'label' to 'labels' for the Trainer
    tokenized_datasets = tokenized_datasets.rename_column("label",
    "labels")
    # Remove original text column
    tokenized_datasets = tokenized_datasets.remove_columns(["text"])
    tokenized_datasets.set_format("torch")
    print("Tokenization complete. Sample:")
    print(tokenized_datasets["train"][0])
except Exception as e:
    print(f"Error during tokenization: {e}")
    exit()

# --- 4. Data Collator ---
data_collator = DataCollatorWithPadding(tokenizer=tokenizer)

# --- 5. Load Model ---
print(f"\nLoading model for sequence classification: {MODEL_CHECKPOINT}")
try:
    model = AutoModelForSequenceClassification.from_pretrained(
        MODEL_CHECKPOINT,
        num_labels=num_labels
        # Optionally pass id_to_label, label_to_id if needed later
    )
    print(f"Model loaded. Initial device: {model.device}")
except Exception as e:
    print(f"Error loading model: {e}")
    exit()

# --- 6. Evaluation Metric ---
print("\nLoading evaluation metric: accuracy")
try:
    accuracy_metric = evaluate.load("accuracy")
    # Could also load F1: f1_metric = evaluate.load("f1")
except Exception as e:
    print(f"Error loading metric: {e}")
    exit()
```

```python
def compute_metrics(eval_pred):
    predictions, labels = eval_pred
    predictions = np.argmax(predictions, axis=1)
    acc = accuracy_metric.compute(predictions=predictions,
    references=labels)
    # Example for adding F1 (requires label count for average)
    # f1 = f1_metric.compute(predictions=predictions, references=labels,
      average='macro') # or 'weighted'
    # return {"accuracy": acc["accuracy"], "f1": f1["f1"]}
    return acc # Just return accuracy for simplicity here

# --- 7. Training Arguments ---
print("\nDefining Training Arguments...")
training_args = TrainingArguments(
    output_dir=OUTPUT_DIR,
    eval_strategy="epoch",
    num_train_epochs=NUM_EPOCHS,
    per_device_train_batch_size=BATCH_SIZE,
    per_device_eval_batch_size=BATCH_SIZE,
    learning_rate=LEARNING_RATE,
    weight_decay=0.01,
    save_strategy="epoch",
    logging_strategy="steps",
    logging_steps=100,
    load_best_model_at_end=True,
    metric_for_best_model="accuracy",
    fp16=torch.cuda.is_available(),
    push_to_hub=False,
    report_to="none"
)
print(f"Training on device: {training_args.device}")

# --- 8. Initialize Trainer ---
trainer = Trainer(
    model=model,
    args=training_args,
    train_dataset=tokenized_datasets["train"],
```

```python
        eval_dataset=tokenized_datasets["validation"],
        tokenizer=tokenizer,
        data_collator=data_collator,
        compute_metrics=compute_metrics,
)

# --- 9. Train ---
print("\nStarting training...")
try:
        train_result = trainer.train()
        print("Training finished.")
        metrics = train_result.metrics
        trainer.log_metrics("train", metrics)
        trainer.save_metrics("train", metrics)
        trainer.save_state()
        trainer.save_model(OUTPUT_DIR) # Save final best model
        print(f"Model and training state saved to {OUTPUT_DIR}")
except Exception as e:
        print(f"Error during training: {e}")
        exit()

# --- 10. Evaluate on Test Set ---
print("\nEvaluating on the test set...")Chapter-5 Prompt
Engineering Mastery
try:
        test_results = trainer.evaluate(eval_dataset=tokenized_
        datasets["test"])
        trainer.log_metrics("test", test_results)
        trainer.save_metrics("test", test_results)
        print("Test Set Evaluation Results:")
        print(test_results)
except Exception as e:
        print(f"Error during test set evaluation: {e}")

# --- End of Recipe ---
```

Another common adaptation is for token-level tasks like **Named Entity Recognition (NER)**. This task is unique because it requires a label for each input token, necessitating careful alignment between the original word labels and the subword tokens generated by the tokenizer.

Recipe 2: Token Classification Specialization (NER)

Fine-tuning a model for Named Entity Recognition

Goal: Fine-tune a model for Named Entity Recognition, a token classification task.

This recipe demonstrates the full fine-tuning workflow for Token Classification, specifically Named Entity Recognition (NER), using an encoder-based model like DistilBERT on the CoNLL-2003 dataset. This task is unique because it requires the model to predict a label (e.g., PERSON, ORGANIZATION) for each input token. The most critical step is carefully aligning the original word labels with the tokenizer's subword tokens to ensure the model learns correctly.

The fine-tuning process for this recipe follows these general steps:

1. **Load Data and Labels:** Load the CoNLL-2003 dataset and extract the list of named entity labels.

2. **Load Tokenizer:** Load the correct DistilBERT tokenizer and verify it is a "fast" tokenizer, which is required for the offset mapping used in label alignment.

3. **Preprocess and Align Labels:** Tokenize the input words and apply a specialized function to align word-level entity tags with sub-tokens (setting labels for special tokens and their subsequent sub-tokens to -100 so they are ignored in the loss calculation).

4. **Data Collator:** Use the dedicated DataCollatorForTokenClassification to handle dynamic padding.

5. **Load Model:** Load the pre-trained model with AutoModelForTokenClassification, which adds a classification head for NER labels.

6. **Evaluation Metric:** Use seqeval, the standard for NER tasks, which reports comprehensive metrics such as F1 score, Precision, and Recall.

7. **Execution and Evaluation:** Initialize and run the Hugging Face Trainer to perform full fine-tuning.

What to Expect

After training for a few epochs (typically 2–3), a fully fine-tuned DistilBERT model on the CoNLL-2003 dataset can be expected to achieve an overall F1 score of approximately 90–91% on the held-out test set, confirming its strong performance on sequence labeling tasks.

```
# --- Recipe: Token Classification Specialization (NER) ---
# Goal: Full fine-tuning of a pre-trained model for Token Classification
(Named Entity Recognition).
# Libraries: transformers, datasets, evaluate, seqeval, torch
# Note: Ensure libraries are installed: pip install transformers datasets
evaluate seqeval accelerate torch
#        Uses DistilBERT and CoNLL-2003 dataset. Requires careful label
         alignment.

import torch
import evaluate
import numpy as np
from datasets import load_dataset, ClassLabel
from transformers import (
    AutoTokenizer,
    AutoModelForTokenClassification,
    TrainingArguments,
    Trainer,
    DataCollatorForTokenClassification # Specific data collator for
    token tasks
)
# --- Configuration ---
MODEL_CHECKPOINT = "distilbert-base-uncased"
DATASET_NAME = "conll2003" # Common NER dataset
NUM_EPOCHS = 1 # Use 1 epoch for quick demo
BATCH_SIZE = 16
LEARNING_RATE = 2e-5
OUTPUT_DIR = "./fine_tune_ner_output"
```

```python
# Use a task name prefix for potential model hub uploads (optional)
TASK_NAME = "ner"
LABEL_ALL_TOKENS = True # Whether to align labels for all sub-tokens or
just the first

# --- 1. Load Data & Labels ---
print(f"Loading dataset: {DATASET_NAME}")
try:
    raw_datasets = load_dataset(DATASET_NAME)
    print("Dataset loaded:")
    print(raw_datasets)

    # Extract label list and mappings from dataset features
    ner_features = raw_datasets["train"].features["ner_tags"]
    label_list = ner_features.feature.names
    label_to_id = {label: i for i, label in enumerate(label_list)}
    id_to_label = {i: label for i, label in enumerate(label_list)}
    num_labels = len(label_list)

    print(f"\nLabels: {label_list}")
    print(f"Number of labels: {num_labels}")
    print(f"\nExample tokens: {raw_datasets['train'][0]['tokens']}")
    print(f"Example NER tags (IDs): {raw_datasets['train'][0]['ner_
tags']}")
    print(f"Example NER tags (Labels): {[id_to_label[tag_id] for tag_id in
raw_datasets['train'][0]['ner_tags']]}")

except Exception as e:
    print(f"Error loading dataset or extracting labels: {e}")
    exit()

# --- 2. Load Tokenizer ---
print(f"\nLoading tokenizer for checkpoint: {MODEL_CHECKPOINT}")
try:
    tokenizer = AutoTokenizer.from_pretrained(MODEL_CHECKPOINT)
    # Check if the tokenizer is fast (required for offset mapping used in
        label alignment)
```

```python
    if not tokenizer.is_fast:
        raise TypeError(
            "This recipe requires a fast tokenizer for offset mapping. "
            f"Try using {MODEL_CHECKPOINT} without '-base-uncased' or
            choose another fast tokenizer."
        )
except Exception as e:
    print(f"Error loading tokenizer: {e}")
    exit()

# --- 3. Preprocess Data (Tokenize and Align Labels) ---
# Function to tokenize inputs and align labels with sub-tokens
def tokenize_and_align_labels(examples):
    tokenized_inputs = tokenizer(
        examples["tokens"],
        truncation=True,
        is_split_into_words=True, # Important: Input is already split
        into words
        padding=False # Padding handled by collator
    )

    labels = []
    for i, label in enumerate(examples[f"{TASK_NAME}_tags"]):
        word_ids = tokenized_inputs.word_ids(batch_index=i) # Map tokens to
        original words
        previous_word_idx = None
        label_ids = []
        for word_idx in word_ids:
            # Special tokens have a word id that is None. We set the label
              to -100 so they are automatically
            # ignored in the loss function.
            if word_idx is None:
                label_ids.append(-100)
            # We set the label for the first token of each word.
            elif word_idx != previous_word_idx:
                label_ids.append(label[word_idx])
```

```
                # For subsequent tokens in a multi-token word, set label based
                  on LABEL_ALL_TOKENS flag
                else:
                    label_ids.append(label[word_idx] if LABEL_ALL_TOKENS
                    else -100)
                previous_word_idx = word_idx

        labels.append(label_ids)

    tokenized_inputs["labels"] = labels
    return tokenized_inputs

print("\nTokenizing dataset and aligning labels...")
try:
    tokenized_datasets = raw_datasets.map(tokenize_and_align_labels,
    batched=True)
    # Remove original columns no longer needed
    tokenized_datasets = tokenized_datasets.remove_columns(raw_
    datasets["train"].column_names)
    print("Tokenization and alignment complete. Sample:")
    sample = tokenized_datasets["train"][0]
    print(f"Input IDs: {sample['input_ids']}")
    print(f"Labels: {sample['labels']}")
    # Decode to see alignment (special tokens have label -100)
    for token_id, label_id in zip(sample['input_ids'], sample['labels']):
            if label_id != -100:
                print(f"{tokenizer.decode([token_id])}: {id_to_label.
                get(label_id, 'N/A')}")
            else:
                print(f"{tokenizer.decode([token_id])}: IGNORED")

except Exception as e:
    print(f"Error during tokenization/alignment: {e}")
    exit()

# --- 4. Data Collator ---
# Handles dynamic padding for token classification tasks
data_collator = DataCollatorForTokenClassification(tokenizer=tokenizer)
```

```python
# --- 5. Load Model ---
print(f"\nLoading model for Token Classification: {MODEL_CHECKPOINT}")
try:
    model = AutoModelForTokenClassification.from_pretrained(
        MODEL_CHECKPOINT,
        num_labels=num_labels,
        id2label=id_to_label, # Pass mappings for better inference/display
        label2id=label_to_id
    )
    print(f"Model loaded. Initial device: {model.device}")
except Exception as e:
    print(f"Error loading model: {e}")
    exit()

# --- 6. Evaluation Metric (Seqeval) ---
print("\nLoading evaluation metric: seqeval (for NER)")
try:
    seqeval_metric = evaluate.load("seqeval")
except Exception as e:
    print(f"Error loading metric: {e}")
    exit()

# Function to compute metrics using seqeval
def compute_metrics(p):
    predictions, labels = p
    predictions = np.argmax(predictions, axis=2) # Get most likely label ID
    for each token

    # Remove ignored index (-100) and convert IDs to label strings
    true_predictions = [
        [label_list[p] for (p, l) in zip(prediction, label) if l != -100]
        for prediction, label in zip(predictions, labels)
    ]
    true_labels = [
        [label_list[l] for (p, l) in zip(prediction, label) if l != -100]
        for prediction, label in zip(predictions, labels)
    ]
```

```python
    results = seqeval_metric.compute(predictions=true_predictions,
    references=true_labels)
    # Return main metrics, can add breakdown per class if needed
    return {
        "precision": results["overall_precision"],
        "recall": results["overall_recall"],
        "f1": results["overall_f1"],
        "accuracy": results["overall_accuracy"],
    }

# --- 7. Training Arguments ---
print("\nDefining Training Arguments...")
training_args = TrainingArguments(
    output_dir=OUTPUT_DIR,
    num_train_epochs=NUM_EPOCHS,
    per_device_train_batch_size=BATCH_SIZE,
    per_device_eval_batch_size=BATCH_SIZE,
    learning_rate=LEARNING_RATE,
    weight_decay=0.01,
    eval_strategy="epoch",
    save_strategy="epoch",
    logging_strategy="steps",
    logging_steps=50,
    load_best_model_at_end=True,
    metric_for_best_model="f1", # Use F1 score for NER typically
    fp16=torch.cuda.is_available(),
    push_to_hub=False,
    report_to="none"
)
print(f"Training on device: {training_args.device}")

# --- 8. Initialize Trainer ---
trainer = Trainer(
    model=model,
    args=training_args,
```

```python
    train_dataset=tokenized_datasets["train"],
    eval_dataset=tokenized_datasets["validation"],
    tokenizer=tokenizer,
    data_collator=data_collator,
    compute_metrics=compute_metrics,
)

# --- 9. Train ---
print("\nStarting training...")
try:
    train_result = trainer.train()
    print("Training finished.")
    metrics = train_result.metrics
    trainer.log_metrics("train", metrics)
    trainer.save_metrics("train", metrics)
    trainer.save_state()
    trainer.save_model(OUTPUT_DIR)
    print(f"Model and training state saved to {OUTPUT_DIR}")
except Exception as e:
    print(f"Error during training: {e}")
    exit()

# --- 10. Evaluate on Test Set ---
print("\nEvaluating on the test set...")
try:
    test_results = trainer.evaluate(eval_dataset=tokenized_
    datasets["test"])
    trainer.log_metrics("test", test_results)
    trainer.save_metrics("test", test_results)
    print("Test Set Evaluation Results:")
    print(test_results)
except Exception as e:
    print(f"Error during test set evaluation: {e}")

# --- End of Recipe ---
```

Adapting Fine-Tuning Across Architectures and Tasks

The fundamental fine-tuning workflow is highly adaptable. While encoder-based models like DistilBERT are a natural fit for classification, the same task can be approached with a decoder-only model. Modern, powerful decoder models (like Gemma or Llama) are so generally capable that, when equipped with a classification head, they can often achieve state-of-the-art performance on NLU tasks. This comes at the cost of higher computational requirements, necessitating techniques like mixed precision training (bf16), gradient accumulation, and gradient checkpointing to fit the process on available hardware.

The next procedure demonstrates how to adapt the classification workflow for a larger, decoder-only model.

Recipe 3: Multi-class Classification (Gemma-2B)

Fine-tuning the decoder-based Gemma-2B SLM on the AG News dataset, adapting it for classification

Goal: Fully fine-tune the decoder-based Gemma-2B model for multi-class text classification on the AG News dataset, highlighting adjustments needed for larger models.

This recipe demonstrates how to adapt the full fine-tuning workflow for a larger, decoder-only model, using the Gemma-2B Small Language Model (SLM) for multi-class classification on the AG News dataset. This is a common and advanced use case, as modern decoder-only models, while computationally intensive, can achieve state-of-the-art performance on Natural Language Understanding (NLU) tasks when properly configured. Key adjustments are necessary to manage computational resources, including reducing batch size, enabling gradient accumulation, using mixed precision training (e.g., bfloat16), and lowering the learning rate.

The fine-tuning process is broken down into the following steps:

1. **Data Preparation:** Load the AG News dataset and create a standard training, validation, and test split.

2. **Tokenizer Configuration:** Load the Gemma-2B tokenizer and ensure its padding is correctly configured, typically by setting the pad token to the end-of-sequence (eos) token for a decoder model.

3. **Model and Training Setup:** Load the model with the AutoModelForSequenceClassification head and use mixed precision (bf16) for memory efficiency if supported by the GPU.

4. **Resource Management**: Define the TrainingArguments, explicitly setting a small batch size, increasing gradient_accumulation_ steps to simulate a larger batch, and enabling memory-saving techniques such as gradient_checkpointing.

5. **Execution and Evaluation:** Initialize the Hugging Face Trainer and run the full fine-tuning process, with evaluation metrics (e.g., accuracy) computed on the validation set after each epoch.

What to Expect

After training for a few epochs (typically 2–3), a fully fine-tuned Gemma-2B model on the AG News dataset can be expected to achieve an accuracy of approximately 94-95% on the held-out test set, demonstrating performance comparable to, or slightly exceeding, smaller encoder models like DistilBERT, but at a higher computational cost.

```python
# --- Recipe: Multi-Class Classification (AG News / Gemma-2B) ---
# Goal: Full fine-tuning of Gemma-2B (a decoder-only SLM) for multi-class
text classification.
# Dataset: AG News (4 classes: World, Sports, Business, Sci/Tech)
# Libraries: transformers, datasets, evaluate, torch, accelerate
# Note: Ensure libraries are installed: pip install transformers datasets
evaluate accelerate torch sentencepiece
#       Fine-tuning Gemma-2B requires significant GPU memory (>16GB VRAM
        recommended, may need adjustments).

import torch
import evaluate
import numpy as np
from datasets import load_dataset
from transformers import (
    AutoTokenizer,
    AutoModelForSequenceClassification,
    TrainingArguments,
```

```python
    Trainer,
    DataCollatorWithPadding
)

# --- Configuration ---
MODEL_CHECKPOINT = "google/gemma-2b" # Using Gemma 2B base model
DATASET_NAME = "ag_news"
NUM_EPOCHS = 1 # Use 1 epoch for quick demo
# Gemma-2B is larger, requires smaller batch size or gradient accumulation
BATCH_SIZE = 4 # Reduced batch size
GRADIENT_ACCUMULATION_STEPS = 4 # Simulate batch size of 4*4=16
LEARNING_RATE = 1e-5 # Often need smaller LR for larger models
OUTPUT_DIR = "./fine_tune_gemma_agnews_output" # Directory for this model

# Check GPU availability and VRAM

if torch.cuda.is_available():
    print(f"GPU detected: {torch.cuda.get_device_name(0)}")
    Simple VRAM check (may not be perfectly accurate)
    t = torch.cuda.get_device_properties(0).total_memory
    r = torch.cuda.memory_reserved(0)
    a = torch.cuda.memory_allocated(0)
    free = t - (r + a) # free Gpu memory
    print(f"Total VRAM: {t / 1024**3:.2f} GB")
    print(f"Approx Free VRAM: {free / 1024**3:.2f} GB")
    if free / 1024**3 < 14: # Rough estimate for Gemma 2B FT
        print("Warning: Low VRAM detected, fine-tuning Gemma-2B
        might fail.")
        print("Consider using QLoRA (Chapter 8) or a smaller model.")
else:
    print("Warning: No GPU detected. Fine-tuning Gemma-2B on CPU will be
    extremely slow.")

import os
os.environ["PYTORCH_CUDA_ALLOC_CONF"] = "expandable_segments:True"
import torch
```

```python
# --- 1. Load Data ---
print(f"Loading dataset: {DATASET_NAME}")
try:
    raw_datasets = load_dataset(DATASET_NAME)
    train_val_split = raw_datasets["train"].train_test_split(test_size=0.1,
    seed=42)
    raw_datasets["train"] = train_val_split["train"]
    raw_datasets["validation"] = train_val_split["test"]
    #print("Dataset loaded and split:")
    #print(raw_datasets[:2])
    label_feature = raw_datasets["train"].features["label"]
    num_labels = label_feature.num_classes
    id_to_label = {i: label for i, label in enumerate(label_feature.names)}
    print(f"Number of labels: {num_labels}")
    print(f"Label mapping: {id_to_label}")
except Exception as e:
    print(f"Error loading dataset: {e}")
    exit()

# --- 2. Load Tokenizer ---
print(f"\nLoading tokenizer for checkpoint: {MODEL_CHECKPOINT}")
try:
    tokenizer = AutoTokenizer.from_pretrained(MODEL_CHECKPOINT)
    # Gemma tokenizer might use <pad> ID 0, but explicitly setting can
      avoid issues
    if tokenizer.pad_token is None:
        print("Setting pad_token to eos_token")
        tokenizer.pad_token = tokenizer.eos_token
        # Important: Resize model embeddings if adding new special tokens
          AFTER model load
        # model.resize_token_embeddings(len(tokenizer))
except Exception as e:
    print(f"Error loading tokenizer: {e}")
    exit()
```

```python
# --- 3. Preprocess Data ---
def tokenize_function(examples):
    # Max length depends on model, Gemma typically supports 2048 or more,
      but keep shorter for classification
    return tokenizer(examples["text"], truncation=True, padding=False, max_
    length=512)

print("\nTokenizing dataset...")
try:
    tokenized_datasets = raw_datasets.map(tokenize_function, batched=True)
    tokenized_datasets = tokenized_datasets.rename_column("label",
    "labels")
    tokenized_datasets = tokenized_datasets.remove_columns(["text"])
    tokenized_datasets.set_format("torch")
    print("Tokenization complete.")
except Exception as e:
    print(f"Error during tokenization: {e}")
    exit()

# --- 4. Data Collator ---
# Use standard padding collator
data_collator = DataCollatorWithPadding(tokenizer=tokenizer)

# --- 5. Load Model ---
print(f"\nLoading model for sequence classification: {MODEL_CHECKPOINT}")
print("This may take a while and require significant RAM/VRAM...")
try:
    model = AutoModelForSequenceClassification.from_pretrained(
        MODEL_CHECKPOINT,
        num_labels=num_labels,
        # Add pad_token_id if you explicitly set tokenizer.pad_token =
          tokenizer.eos_token
        # pad_token_id=tokenizer.pad_token_id,
        torch_dtype=torch.bfloat16 if torch.cuda.is_available() and torch.
        cuda.is_bf16_supported() else torch.float32 # Use bfloat16 if
        possible
    )
```

```python
    # If PAD token was added/changed AFTER loading tokenizer but BEFORE
      model, resize needed:
    # model.resize_token_embeddings(len(tokenizer))
    # Explicitly set pad_token_id in model config if it wasn't set
      correctly
    if model.config.pad_token_id is None:
        model.config.pad_token_id = tokenizer.pad_token_id
        print(f"Set model.config.pad_token_id to: {model.config.pad_
        token_id}")

    print(f"Model loaded. Initial device: {model.device}") # Trainer
    will move it
except Exception as e:
    print(f"Error loading model: {e}")
    print("Check VRAM, RAM, and potential Hugging Face Hub connectivity
    issues.")
    exit()

# --- 6. Evaluation Metric ---
print("\nLoading evaluation metric: accuracy")
try:
    accuracy_metric = evaluate.load("accuracy")
except Exception as e:
    print(f"Error loading metric: {e}")
    exit()

def compute_metrics(eval_pred):
    predictions, labels = eval_pred
    predictions = np.argmax(predictions, axis=1)
    return accuracy_metric.compute(predictions=predictions,
    references=labels)

# --- 7. Training Arguments ---
print("\nDefining Training Arguments...")
# Check for bfloat16 support for mixed precision
bf16_supported = torch.cuda.is_available() and torch.cuda.is_bf16_
supported()
```

```python
training_args = TrainingArguments(
    output_dir=OUTPUT_DIR,
    num_train_epochs=NUM_EPOCHS,
    per_device_train_batch_size=BATCH_SIZE,
    per_device_eval_batch_size=BATCH_SIZE * 2, # Can often use larger eval
    batch size
    gradient_accumulation_steps=GRADIENT_ACCUMULATION_STEPS,
    learning_rate=LEARNING_RATE,
    weight_decay=0.01,
    eval_strategy="epoch",
    save_strategy="epoch",
    logging_strategy="steps",
    logging_steps=50, # Log more frequently with accumulation
    load_best_model_at_end=True,
    metric_for_best_model="accuracy",
    fp16=torch.cuda.is_available() and not bf16_supported, # Use fp16 if
    bf16 not supported
    bf16=bf16_supported, # Use bf16 if supported (preferred for newer GPUs)
    gradient_checkpointing=True, # Saves memory at cost of slower training
    push_to_hub=False,
    report_to="none"
)
print(f"Training on device: {training_args.device}")
print(f"Using Gradient Accumulation: {GRADIENT_ACCUMULATION_STEPS} steps")
print(f"Effective Batch Size: {BATCH_SIZE * GRADIENT_ACCUMULATION_STEPS}")
print(f"Mixed precision (fp16): {training_args.fp16}")
print(f"Mixed precision (bf16): {training_args.bf16}")

# --- 8. Initialize Trainer ---
trainer = Trainer(
    model=model,
    args=training_args,
    train_dataset=tokenized_datasets["train"],
    eval_dataset=tokenized_datasets["validation"],
    tokenizer=tokenizer,
    data_collator=data_collator,
```

```python
    compute_metrics=compute_metrics,
)

# --- 9. Train ---
print("\nStarting training (Gemma-2B)...")
try:
    train_result = trainer.train()
    print("Training finished.")
    metrics = train_result.metrics
    trainer.log_metrics("train", metrics)
    trainer.save_metrics("train", metrics)
    trainer.save_state()
    trainer.save_model(OUTPUT_DIR) # Save final best model
    print(f"Model and training state saved to {OUTPUT_DIR}")
except Exception as e:
    print(f"Error during training: {e}")
    print("This might be an Out-of-Memory (OOM) error. Try reducing
    batch size, enabling gradient checkpointing, or using LoRA/QLoRA
    (Chapter 8).")
    exit()

# --- 10. Evaluate on Test Set ---
print("\nEvaluating on the test set...")
try:
    test_results = trainer.evaluate(eval_dataset=tokenized_
    datasets["test"])
    trainer.log_metrics("test", test_results)
    trainer.save_metrics("test", test_results)
    print("Test Set Evaluation Results:")
    print(test_results)
except Exception as e:
    print(f"Error during test set evaluation: {e}")

# --- End of Recipe ---
```

Generative Fine-Tuning

Beyond classification, the fine-tuning paradigm can be shifted to entirely different objectives. **Generative fine-tuning**, for example, does not aim to predict a class label but to adapt the model's writing style or knowledge domain. This requires a different model head (AutoModelForCausalLM) and a data preparation strategy that structures the text for next-token prediction.

Recipe 4: Generative Adaptation (Causal LM)

Fine-tuning a model like DistilGPT-2 to generate text in a specific style (e.g., adapting to a particular author's voice).

Goal: Fine-tune a causal language model (like DistilGPT-2) to adapt its text generation style or knowledge domain using the Wikitext dataset as a demonstration.

Beyond classification, the fine-tuning paradigm can be shifted to an entirely different objective: Generative Fine-Tuning. This process does not aim to predict a class label but instead adapts the model's writing style or knowledge domain. It requires a different model head (AutoModelForCausalLM) and a data preparation strategy that structures the text for next-token prediction, which is the model's primary training objective. This recipe uses the DistilGPT-2 model to adapt its style using the Wikitext-2 dataset.

The fine-tuning workflow for this recipe follows these general steps:

1. **Load Dataset Subsets:** Load the necessary splits of the Wikitext dataset.

2. **Load Tokenizer:** Load the correct tokenizer for the model (e.g., DistilGPT-2) and ensure the padding token is correctly configured, typically by setting it to the end-of-sequence (eos) token for a decoder model.

3. **Tokenization:** Convert the text examples into numerical IDs.

4. **Group Tokens into Blocks:** Concatenate all tokenized texts and then split them into fixed length blocks (e.g., 128 tokens) to ensure consistent batching and efficient training for the Causal LM objective.

5. **Load Model:** Load the pre-trained model using AutoModelForCausalLM.

6. **Data Collator:** Use DataCollatorForLanguageModeling with mlm=False to prepare the batches for the causal language modeling objective (next-token prediction).

7. **Trainer Setup:** Initialize the Hugging Face Trainer with appropriate training arguments.

8. **Train and Evaluate:** Execute the training process and calculate the final perplexity on the validation set.

What to Expect

After training for the configured number of epochs, the key metric to observe will be the perplexity value, which is derived from the final validation loss. A successful fine-tuning run is indicated by a perplexity value substantially lower than the model's initial perplexity on the dataset, demonstrating a successful adaptation of the model's generation capabilities to the target domain or style.

```python
# --- Recipe: Causal LM Fine-Tuning on Wikitext-2 ---
# Goal: Fine-tune GPT-2 on Wikitext-2 with tokenization and block grouping

import torch
import math
from datasets import load_dataset
from transformers import (
    AutoTokenizer,
    AutoModelForCausalLM,
    Trainer,
    TrainingArguments,
    DataCollatorForLanguageModeling,
    pipeline,
)

# --- Configuration ---
MODEL_CHECKPOINT = "distilgpt2"
DATASET_NAME = "wikitext"
DATASET_CONFIG = "wikitext-2-raw-v1"
NUM_EPOCHS = 1
BATCH_SIZE = 4
LEARNING_RATE = 2e-5
```

```python
BLOCK_SIZE = 128
OUTPUT_DIR = "./fine_tuned_gpt2_wikitext"

# --- 1. Load Dataset Subsets ---
print("Loading Wikitext dataset...")
train_dataset = load_dataset(DATASET_NAME, DATASET_CONFIG,
split="train[:5000]")
val_dataset = load_dataset(DATASET_NAME, DATASET_CONFIG,
split="validation[:500]")

# --- 2. Load Tokenizer ---
tokenizer = AutoTokenizer.from_pretrained(MODEL_CHECKPOINT)
if tokenizer.pad_token is None:
    tokenizer.pad_token = tokenizer.eos_token
    print(f"Set PAD token to EOS token: {tokenizer.pad_token}")

# --- 3. Tokenization ---
def tokenize_function(examples):
    return tokenizer(examples["text"])

tokenized_train = train_dataset.map(tokenize_function, batched=True,
remove_columns=["text"])
tokenized_val = val_dataset.map(tokenize_function, batched=True, remove_
columns=["text"])

# --- 4. Group Tokens into Blocks ---
def group_texts(examples):
    # Concatenate all examples for each key
    concatenated = {k: sum(examples[k], []) for k in examples.keys()}
    total_length = len(concatenated["input_ids"])

    # Truncate to a multiple of block_size
    if total_length >= BLOCK_SIZE:
        total_length = (total_length // BLOCK_SIZE) * BLOCK_SIZE

    # Split into blocks
    result = {
        k: [concatenated[k][i : i + BLOCK_SIZE] for i in range(0, total_
        length, BLOCK_SIZE)]
```

```python
        for k in concatenated.keys()
    }

    return result

lm_train_dataset = tokenized_train.map(group_texts, batched=True)
lm_val_dataset = tokenized_val.map(group_texts, batched=True)

# --- 5. Load Model ---
model = AutoModelForCausalLM.from_pretrained(MODEL_CHECKPOINT)

# --- 6. Data Collator ---
data_collator = DataCollatorForLanguageModeling(tokenizer=tokenizer,
mlm=False)

# --- 7. Training Arguments ---
training_args = TrainingArguments(
    output_dir=OUTPUT_DIR,
    eval_strategy="epoch",
    save_strategy="epoch",
    learning_rate=LEARNING_RATE,
    per_device_train_batch_size=BATCH_SIZE,
    per_device_eval_batch_size=BATCH_SIZE,
    num_train_epochs=NUM_EPOCHS,
    weight_decay=0.01,
    logging_steps=50,
    push_to_hub=False,
    load_best_model_at_end=True,
    metric_for_best_model="eval_loss",
    fp16=torch.cuda.is_available(),
    report_to="none",
)

# --- 8. Trainer ---
trainer = Trainer(
    model=model,
    args=training_args,
    train_dataset=lm_train_dataset,
```

```
    eval_dataset=lm_val_dataset,
    tokenizer=tokenizer,
    data_collator=data_collator,
)

# --- 9. Train ---
print("\nTraining...")
trainer.train()
trainer.save_model(OUTPUT_DIR)

# --- 10. Evaluate (Perplexity) ---
print("\nEvaluating model...")
eval_results = trainer.evaluate()
perplexity = math.exp(eval_results["eval_loss"])
print(f"Perplexity: {perplexity:.2f}")

# --- 11. Generate Text (Optional) ---
print("\nGenerating text...")
generator = pipeline("text-generation", model=OUTPUT_DIR,
tokenizer=tokenizer, device=0 if torch.cuda.is_available() else -1)
prompt = "Artificial intelligence is"
output = generator(prompt, max_new_tokens=50, num_return_sequences=1)[0]
["generated_text"]

print(f"Prompt: {prompt}\nGenerated: {output}")
```

Experiment Tracking and Comparative Analysis

Given the iterative nature of fine-tuning and the number of variables involved (model choice, hyperparameters, and data), systematically tracking experiments is crucial for reproducibility and analysis. Furthermore, directly comparing the performance of different fine-tuned models on a standardized test set provides invaluable, objective insights for making final architectural and modeling decisions.

Recipe 5: Logging with Weights and Biases

Integrating Weights & Biases (W&B) for comprehensive experiment tracking and visualization (applicable to any training recipe)

Goal: Integrate the Weights and Biases (W&B) platform into the fine-tuning workflow to enable robust experiment tracking, real-time metric visualization, and automatic artifact (model/data) management.

This recipe focuses on adopting a best-practice MLOps tool for professional-grade experiment management. The Hugging Face Trainer is designed to seamlessly integrate with W&B, allowing for automatic logging of all training, validation, and test metrics, as well as the saving of model configurations and final checkpoints. This is essential for comparing different hyperparameter choices, debugging training runs, and ensuring reproducibility.

The fine-tuning workflow integration follows these general steps:

1. **Installation and Login:** Ensure the W&B library is installed and authenticated either via the command line or an environment variable/notebook login command.

2. **Environment Configuration:** Set the report_to parameter in TrainingArguments to include "wandb" to activate the integration. Optionally, set run_name for clear identification.

3. **Automatic Logging:** The Trainer will automatically log the training loss, learning rate, evaluation metrics (like accuracy or F1), and system metrics (GPU utilization and memory).

4. **Artifact Management:** W&B automatically saves the final model and configuration as an artifact, enabling easy version control and deployment.

5. **Run Visualization:** All logged metrics are pushed to the W&B dashboard, allowing for real-time visualization, comparison with other runs, and table creation of model performance.

```
pip install wandb
# --- Configuration (Same as Classification Recipe) ---
MODEL_CHECKPOINT = "distilbert-base-uncased"
DATASET_NAME = "imdb"
```

```python
NUM_EPOCHS = 1
BATCH_SIZE = 16
LEARNING_RATE = 2e-5
OUTPUT_DIR = "./fine_tune_classifier_wandb_output" # Use a different
output dir

# --- Steps 1-6 (Load Data, Tokenizer, Preprocess, Collator, Model,
Metrics) ---
# These steps are identical to the 'ch6_recipe_classification_ft' recipe.
# For brevity, assume these steps have been executed and we have:
# - tokenized_datasets (containing 'train', 'validation', 'test')
# - tokenizer
# - model
# - data_collator
# - compute_metrics function
# --- Start copy-paste block from ch6_recipe_classification_ft (if running
standalone) ---
print(f"Loading dataset: {DATASET_NAME}")
raw_datasets = load_dataset(DATASET_NAME)
train_val_split = raw_datasets["train"].train_test_split(test_size=0.1,
seed=42)
raw_datasets["train"] = train_val_split["train"]
raw_datasets["validation"] = train_val_split["test"]
print(f"Loading tokenizer for checkpoint: {MODEL_CHECKPOINT}")
tokenizer = AutoTokenizer.from_pretrained(MODEL_CHECKPOINT)
def tokenize_function(examples):
    return tokenizer(examples["text"], truncation=True, padding=False)
print("Tokenizing dataset...")
tokenized_datasets = raw_datasets.map(tokenize_function, batched=True)
tokenized_datasets = tokenized_datasets.remove_columns(["text"])
tokenized_datasets = tokenized_datasets.rename_column("label", "labels")
tokenized_datasets.set_format("torch")
data_collator = DataCollatorWithPadding(tokenizer=tokenizer)
print(f"Loading model for sequence classification: {MODEL_CHECKPOINT}")
num_labels = raw_datasets["train"].features["label"].num_classes
```

```python
model = AutoModelForSequenceClassification.from_pretrained(MODEL_
CHECKPOINT, num_labels=num_labels)
print("Loading evaluation metric: accuracy")
accuracy_metric = evaluate.load("accuracy")
def compute_metrics(eval_pred):
    predictions, labels = eval_pred
    predictions = np.argmax(predictions, axis=1)
    return accuracy_metric.compute(predictions=predictions,
references=labels)
# --- End copy-paste block ---

# --- 7. Training Arguments (Modified for W&B) ---
print("\nDefining Training Arguments with W&B Reporting...")
training_args = TrainingArguments(
    output_dir=OUTPUT_DIR,
    num_train_epochs=NUM_EPOCHS,
    per_device_train_batch_size=BATCH_SIZE,
    per_device_eval_batch_size=BATCH_SIZE,
    learning_rate=LEARNING_RATE,
    weight_decay=0.01,
    eval_strategy="epoch",
    save_strategy="epoch",
    logging_strategy="steps",
    logging_steps=50, # Log more frequently to see updates in W&B
    load_best_model_at_end=True,
    metric_for_best_model="accuracy",
    fp16=torch.cuda.is_available(),
    push_to_hub=False,
    # *** Key Change: Enable W&B reporting ***
    report_to="wandb",
    # Optional: Name your W&B run
    run_name=f"{MODEL_CHECKPOINT}-ft-{DATASET_NAME}-demo"
)
print(f"Training on device: {training_args.device}")
print(f"Reporting to W&B project: {WANDB_PROJECT_NAME}")
```

```python
# --- 8. Initialize Trainer (Same as before) ---
trainer = Trainer(
    model=model,
    args=training_args,
    train_dataset=tokenized_datasets["train"],
    eval_dataset=tokenized_datasets["validation"],
    tokenizer=tokenizer,
    data_collator=data_collator,
    compute_metrics=compute_metrics,
)

# --- 9. Train (W&B logging happens automatically) ---
print("\nStarting training (logging to W&B)...")
# You should see W&B links printed in your console output.
try:
    train_result = trainer.train()
    print("Training finished.")
    metrics = train_result.metrics
    trainer.log_metrics("train", metrics)
    trainer.save_metrics("train", metrics)
    trainer.save_state()
    trainer.save_model(OUTPUT_DIR)
    print(f"Model and training state saved to {OUTPUT_DIR}")
except Exception as e:
    print(f"Error during training: {e}")
    # Ensure W&B run is finished even if there's an error
    import wandb
    wandb.finish()
    exit()

# --- 10. Evaluate on Test Set (Also logged to W&B) ---
print("\nEvaluating on the test set (logging to W&B)...")
try:
    test_results = trainer.evaluate(eval_dataset=tokenized_
    datasets["test"])
    trainer.log_metrics("test", test_results)
```

```
    trainer.save_metrics("test", test_results)
    print("Test Set Evaluation Results:")
    print(test_results)
except Exception as e:
    print(f"Error during test set evaluation: {e}")

# --- 11. Finish W&B Run ---
# Important to properly close the W&B run
import wandb
wandb.finish()
print("\nW&B run finished.")

# --- End of Recipe ---
```

Chapter Summary

We have explored Full-Parameter Fine-Tuning in detail, establishing it as the most
intensive yet powerful method for specializing a pre-trained Large Language Model (LLM).
You saw how transfer learning is the foundational principle, allowing us to efficiently adapt
the vast, general knowledge of a base model to a much smaller, domain-specific dataset.
We discussed the critical trade-offs: while full fine-tuning offers the highest performance
potential by updating every parameter, it also demands substantial resources (especially
VRAM) and carries the risk of catastrophic forgetting, which you must actively mitigate
with techniques such as small learning rates and regularization. We then provided a
decision framework, clarifying that this resource-intensive approach is the right choice
when your domain is significantly different, achieving the absolute maximum performance
is paramount, or you need a strong performance baseline. We walked you through the
systematic Fine-Tuning Workflow, which includes crucial steps like loading the correct
Tokenizer, meticulous Data Preprocessing (such as label alignment for Named Entity
Recognition), and the vital process of Hyperparameter Optimization, where you must
carefully tune the Learning Rate (typically 1e–5 to 5e–5) and Batch Size for success.
Finally, we demonstrated the versatility of the approach across different architectures
and tasks with practical recipes: from encoder-based models like DistilBERT for efficient
Multi-class Classification and Token Classification to adapting the larger, decoder-only

Gemma-2B model for classification using resource-management techniques like gradient accumulation and mixed-precision (bf16) and even fine-tuning for Generative Adaptation to adjust the model's style. Now that you have mastered the traditional method of full fine-tuning, we will next explore a more advanced technique: instruction fine-tuning, where we teach a model not just a specific task but the general skill of following diverse instructions.

Instruction Fine-Tuning

In the previous chapter, we learned how to adapt a base model to a specific domain using full fine-tuning. Now, we will teach the model a more general and powerful skill: how to read, understand, and execute *any* instruction we provide with precision and accuracy. This is the Instruction Fine-Tuning (IFT) process. It is about transforming a base language model—often one trained simply to predict the next word—into a responsive and helpful assistant capable of effectively tackling diverse commands, questions, and prompts, even those it has not encountered before.

What Is Instruction Fine-Tuning?

The primary goal of IFT is **alignment**. We want to align a pre-trained language model (typically a decoder-only Causal LM like Llama, Gemma, or Mistral) with **user intent**. We're teaching it not just to predict text but to understand and follow instructions in a zero-shot way, meaning it can handle new instructions it wasn't explicitly trained on.

This extra step is essential because base Causal Language Models like google/gemma-2b or meta-llama/Llama-3-8b are not inherently good at following instructions. They are powerful **base models**, trained to be expert predictors of the next word. If you give a base model an instruction like "Summarize this paragraph," it's likely to continue the sentence in a way it has seen online, such as "...is a common task in natural language processing," rather than performing the summarization. It only learns this instruction-following behavior after IFT.

The **mechanism** involves fine-tuning the base LLM on a specially curated dataset. This dataset not only contains text but also includes examples that demonstrate *how to follow instructions*. Each example typically includes an `instruction` (the command), optional `input` or context (background information needed), and the desired `output` (the ideal response).

B. K. Bolla et al., *Large Language Model Recipes*, https://doi.org/10.1007/979-8-8688-2607-8_7

How does this **contrast with standard fine-tuning** (Chapter 6)? Standard fine-tuning often optimizes for a specific metric on a relatively narrow task, such as improving accuracy in sentiment classification. IFT, however, aims for broader capabilities—optimizing for helpfulness, safety, accuracy, and the general ability to follow diverse instructions across tasks such as summarization, question answering, brainstorming, coding, and creative writing.

Essentially, **all base Causal Language Models need IFT** to become helpful assistants. When you see a model with a name like gemma-it or Llama-3-Instruct, the -it or -Instruct suffix signals that it has already undergone this crucial IFT process.

Instruction Fine-Tuning (IFT) and subsequent advanced alignment techniques like Reinforcement Learning from Human Feedback (RLHF) or Direct Preference Optimization (DPO) are sequential and essential steps in transforming a base Language Model (LLM) into a helpful and competent assistant.

Instructional Fine-Tuning (IFT) establishes foundational behavior by training a base model on structured examples, teaching it to follow a format and act as an assistant.

> **Goal: Alignment with User Intent.** The primary objective is to bridge the gap between a Causal LM's raw predictive power and its ability to act on explicit commands. A base model is an "expert predictor of the next word" and must be taught to recognize and execute instructions in a zero-shot manner.

> **Mechanism: Structured Data and Loss Masking.** IFT relies on a specialized dataset where each example explicitly contains the instruction, optional input context, and the desired output/response. Crucially, the training employs loss masking to ignore the instruction and input tokens during loss calculation, forcing the model's learning to focus solely on generating the appropriate response. This transforms its behavior from text continuation to task completion.

The subsequent advanced alignment step, **RLHF/Direct Preference Optimization (DPO),** refines the model's values and preferences, making it helpful, safer, and less biased.

Goal: Refinement of Values and Safety. While IFT teaches the model how to respond, it doesn't guarantee the response is optimal. The model may still be biased, unhelpful, or unsafe based on its vast, uncurated pre-training data.

Mechanism: Human Preference Data. This phase adds an extra layer of refinement by training the model on human judgments and comparisons (preferences). This data explicitly guides the model toward more helpful, truthful, and harmless responses. RLHF, for example, uses a Reward Model trained on these preferences to further fine-tune the LLM.

This two-stage process culminates in a competent LLM assistant, signaled by suffixes such as "-it" or "-Instruct," that aligns with human directives and conversational patterns.

Why Is Instruction Fine-Tuning Necessary?

This extra step of instruction fine-tuning is not just a minor enhancement; it is often essential for crafting effective, interactive AI models. It effectively bridges the gap between a model's raw predictive power and its genuine usefulness. Base LLMs, fresh from their extensive pre-training, can be surprisingly literal. If you ask one a question, it might simply continue your sentence as if completing a paragraph it found online, rather than providing a direct answer. IFT explicitly trains the model on expected conversational patterns and task completion, teaching it to recognize an instruction and respond appropriately. Figure 7-1 illustrates the base model's response before and after fine-tuning on the instruction task.

Figure 7-1. *The Impact of Instruction Fine-Tuning on Language Model Behavior. This figure demonstrates how instruction fine-tuning transforms language models from text predictors to instruction followers. Pre-trained models (left) treat prompts as text to continue, while instruction-tuned models (right) interpret the same prompts as tasks to perform*

Furthermore, IFT significantly improves **zero-shot performance** on a wide range of tasks. An instruction-tuned model exhibits significantly improved generalization. Even when presented with an instruction for a task it was not specifically trained on, it has a much greater chance of generating a relevant and useful response than its non-instruction-tuned base version. Ultimately, IFT serves as the foundation for creating effective chatbots, helpful virtual assistants, and applications where the AI is expected to interact naturally and follow user directives.

The success of this process hinges almost entirely on the quality and structure of the data used for training.

The Critical Role of the Instruction Dataset

The success of IFT depends critically on the dataset used for tuning. Its quality, diversity, and scale are paramount. Each example in an instruction dataset typically includes

- **Instruction:** The core task or query (e.g., "Summarize the following text.")

- **Input (Optional):** Extra context needed for the instruction (e.g., the text to be summarized)

- **Output/response:** The high-quality, desired answer the model should generate

Instruction Dataset Sourcing Methodologies

The acquisition of a high-quality instruction dataset can be approached through several established methodologies. The most direct method is leveraging **Publicly Available Datasets**, such as databricks/databricks-dolly-15k, the Stanford Alpaca dataset, and OpenAssistant/oasst1, which serve as excellent starting points. An alternative, highly scalable approach is **Synthetic Data Generation**, often referred to as the Self-Instruct technique. This method utilizes a powerful existing LLM as a generator, which is prompted with a few seed examples to produce a large volume of new instruction-response pairs, though this can risk inheriting the generator model's intrinsic biases. The third methodology, which yields the highest fidelity data, is **Human Annotation**. In this process, human experts write, refine, and verify instructions and their corresponding outputs. While this approach produces optimal quality, it is significantly more resource intensive in terms of cost and time. Regardless of the chosen methodology, several key considerations are paramount during dataset development: the **diversity** of instructions, the **quality** of the responses, the identification and mitigation of potential **biases**, and the execution of a thorough **cleaning** process.

Figure 7-2 illustrates the intuitive view of the structure of instruction fine-tuning datasets.

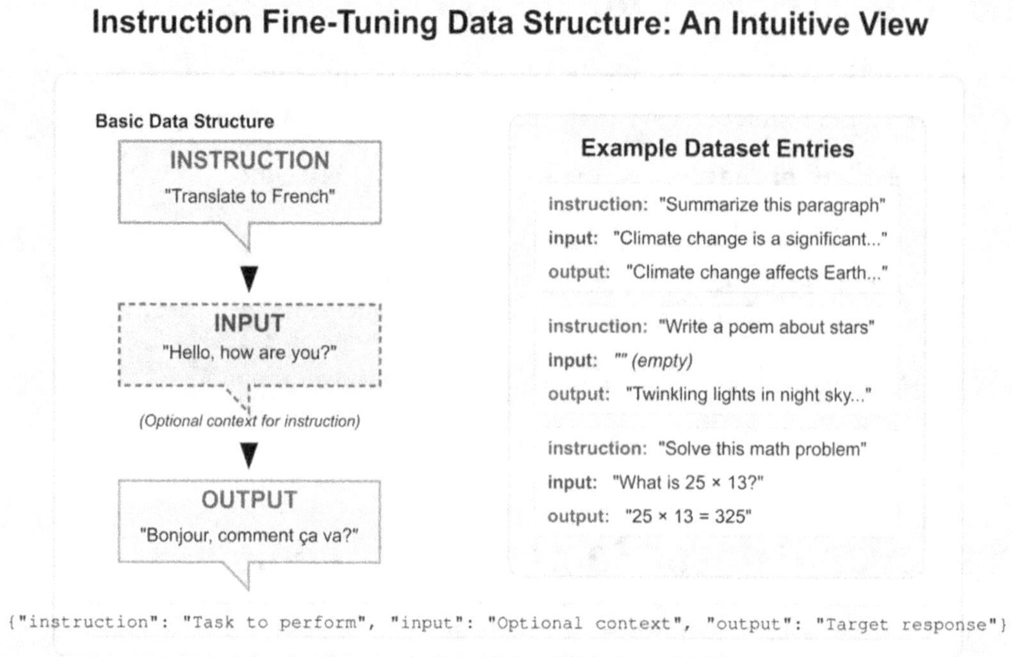

Instruction Fine-Tuning Data Structure: An Intuitive View

Figure 7-2. Instruction Fine-Tuning Data Structure. This diagram illustrates the three-part structure of instruction fine-tuning data: the instruction (what task to perform), optional input (context for the instruction), and output (target response)

Once a high-quality dataset is secured, it must be formatted correctly so the model can learn from it effectively.

Data Formatting and Prompt Templating

Since IFT usually targets Causal LMs (which predict the next token), we need to convert our structured instruction, input, and output data into a single, continuous string of text. This is where **prompt templates** come in. These are predefined structures used consistently during training to format each data example. The model learns to recognize this template, understanding which part is the instruction and which part is the response it needs to generate. Common styles include the Alpaca format, the Vicuna role-based format, and the token-based formats used by models like Gemma and Llama. These different types of dataset structures that are used for instruction fine-tuning are illustrated in Figure 7-3.

Common Instruction Fine-Tuning Template Formats

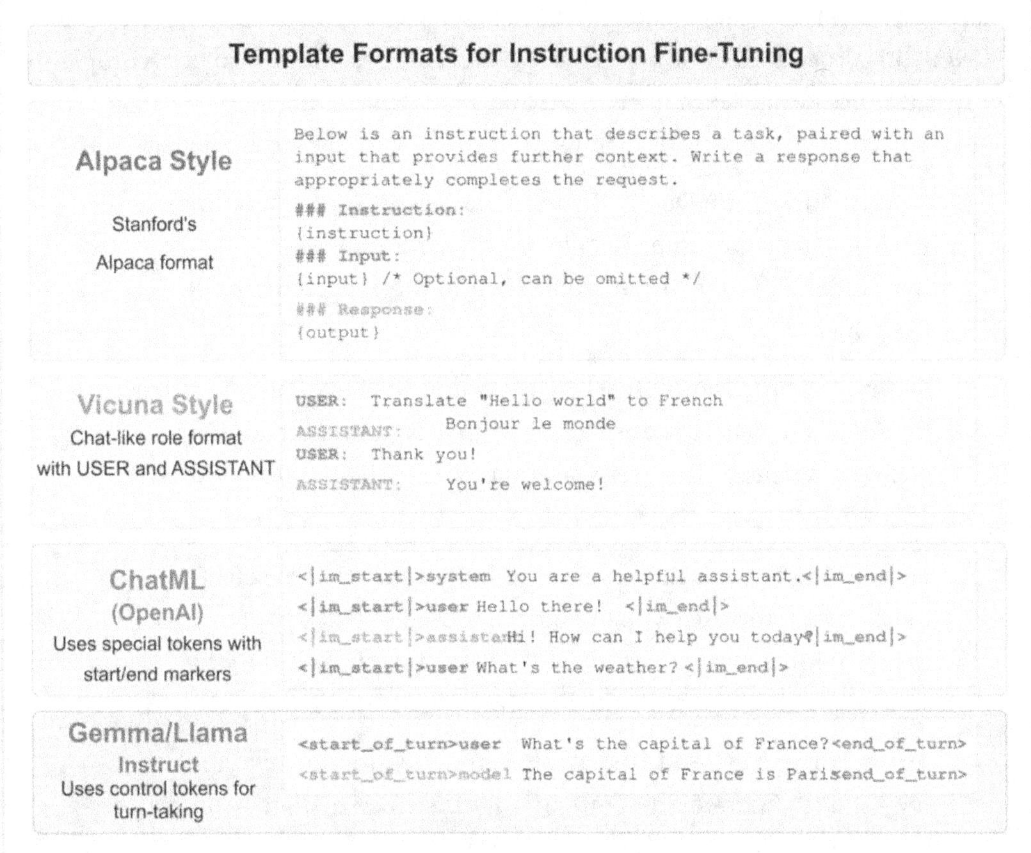

Figure 7-3. *Common Instruction Fine-Tuning Template Formats. This diagram compares four standard formats used to structure instruction tuning data: Alpaca Style, Vicuna Style, ChatML, and Gemma/Llama Instruct*

Recipe 1: Formatting Instructions with a Prompt Template

Applying a prompt template to structure raw instruction data (e.g., from JSON to a formatted string)

Goal: Apply a specific prompt template to structure raw instruction data into a format ready for training.

This recipe demonstrates the critical first step in Instruction Fine-Tuning (IFT): adapting raw, structured instruction data into a single, continuous text format suitable for training a Causal Language Model (CLM). Since CLMs are trained to predict the next token, this procedure establishes the specific conversational and task-completion structure the model must learn to recognize and follow.

The instruction-formatting workflow for this recipe follows these general steps:

1. **Sample Raw Data:** Define the input data as a list of dictionaries, ensuring each example contains the core components: instruction, optional input (context), and the expected output (response).

2. **Define Prompt Templates:** Create predefined text structures (like the Alpaca format) to consistently wrap the instruction and input, clearly delineating the user's command from the model's desired response.

3. **Formatting Function:** Implement a helper function to select the correct template (with or without input context), apply the formatting, and concatenate the resulting prompt with the expected output to create the final training string.

4. **Apply Formatting and Display:** Execute the function over the raw data to generate the sequence of text strings that are ready for tokenization and the subsequent training process.

What to Expect

After applying the template, the output is a continuous string of text for each training example. This string contains the full conversation (instruction, input, and response) in a consistent format. This formatted string is the fundamental input for the next stage, Loss Masking (Recipe 2), as it explicitly teaches the model the exact sequence of tokens it must generate to fulfill a new instruction.

This recipe demonstrates the critical first step in Instruction Fine-Tuning (IFT): adapting raw, structured instruction data into a single, continuous text format suitable for training a Causal Language Model (CLM). Since CLMs are trained to predict the next token, this procedure establishes the specific conversational and task-completion structure the model must learn to recognize and follow.

```
pip install datasets

# --- Recipe: Formatting the Instructions ---
# Goal: Apply a specific prompt template to structure raw instruction data.
# Method: Uses a simple Python function and demonstrates the Alpaca
template.

import json

# --- 1. Sample Raw Data (List of Dictionaries) ---
# Represents data you might load from a file or API
raw_data = [
    {
        "instruction": "Provide a brief summary of the concept of transfer
        learning in machine learning.",
        "input": "", # No additional input needed
        "output": "Transfer learning is a machine learning technique where
        a model developed for a task is reused as the starting point for a
        model on a second, related task. It leverages knowledge gained from
        the source task to improve performance on the target task, often
        reducing the need for large amounts of target-specific data."
    },
    {
        "instruction": "Translate the following sentence to Spanish.",
        "input": "Hello, how are you?",
        "output": "Hola, ¿cómo estás?"
    },
    {
        "instruction": "List three common types of renewable energy
        sources.",
        "input": "",
        "output": "1. Solar Energy\n2. Wind Energy\n3.
        Hydroelectric Energy"
    }
]
```

```python
# --- 2. Define Prompt Templates ---
# Using the Alpaca format as an example

# Template for instructions WITH input context
PROMPT_WITH_INPUT_TEMPLATE = """Below is an instruction that describes a
task, paired with an input that provides further context. Write a response
that appropriately completes the request.

### Instruction:
{instruction}

### Input:
{input}

### Response:
""" # Output will be appended here during training/inference

# Template for instructions WITHOUT input context
PROMPT_NO_INPUT_TEMPLATE = """Below is an instruction that describes a
task. Write a response that appropriately completes the request.

### Instruction:
{instruction}

### Response:
""" # Output will be appended here during training/inference

# --- 3. Formatting Function ---
def format_instruction_data(example):
    """Applies the appropriate Alpaca prompt template."""
    instruction = example.get("instruction", "")
    input_context = example.get("input", "")
    output = example.get("output", "") # Output is needed for
    training labels

    if input_context and input_context.strip():
        # Use the template with input
        prompt_start = PROMPT_WITH_INPUT_TEMPLATE.format(
            instruction=instruction,
            input=input_context
        )
```

```python
    else:
        # Use the template without input
        prompt_start = PROMPT_NO_INPUT_TEMPLATE.format(
            instruction=instruction
        )

    # For training, we concatenate the prompt start and the expected output
    # For inference, we only use prompt_start
    formatted_text_for_training = prompt_start + output

    return {
        "formatted_prompt": prompt_start, # Useful for inference later
        "formatted_training_text": formatted_text_for_training
    }

# --- 4. Apply Formatting ---
print("--- Formatting Raw Data ---")
formatted_data = []
for example in raw_data:
    formatted_example = format_instruction_data(example)
    formatted_data.append(formatted_example)

# --- 5. Display Results ---
for i, item in enumerate(formatted_data):
    print(f"\n--- Example {i+1} ---")
    print(f"Original: {raw_data[i]}")
    print("-" * 20)
    print(f"Formatted Prompt (for Inference):\n{item['formatted_prompt']}")
    print("-" * 20)
    print(f"Formatted Text (for Training):\n{item['formatted_training_
    text']}")
    print("=" * 40)

# --- Notes ---
# - This formatted_training_text is what you would tokenize for Causal LM
fine-tuning.
# - During tokenization for training, you need to identify which tokens
belong to the
```

```
#    'output' part to avoid masking them during loss calculation.
# - Remember to add the EOS token to the end of formatted_training_text
before tokenization.

# --- End of Recipe ---
```

The Training Process: Understanding Loss Masking

The underlying training objective for IFT is still standard Causal Language Modeling: predict the next token. However, there is a critical twist: we only want the model to learn **how to generate the response**. We do not want it learning to predict the instruction or input text. To achieve this, we use **loss masking**. We mask the loss calculation for the tokens corresponding to the instruction and input parts of the template. This means the model's learning focuses solely on predicting the tokens in the desired output section. A typical implementation technique is to set the target labels for the masked tokens to a value (like -100) that is ignored by the loss function. Figure 7-4 illustrates the selective training process, demonstrating how loss masking focuses learning entirely on the desired output tokens while ignoring the instruction and input context.

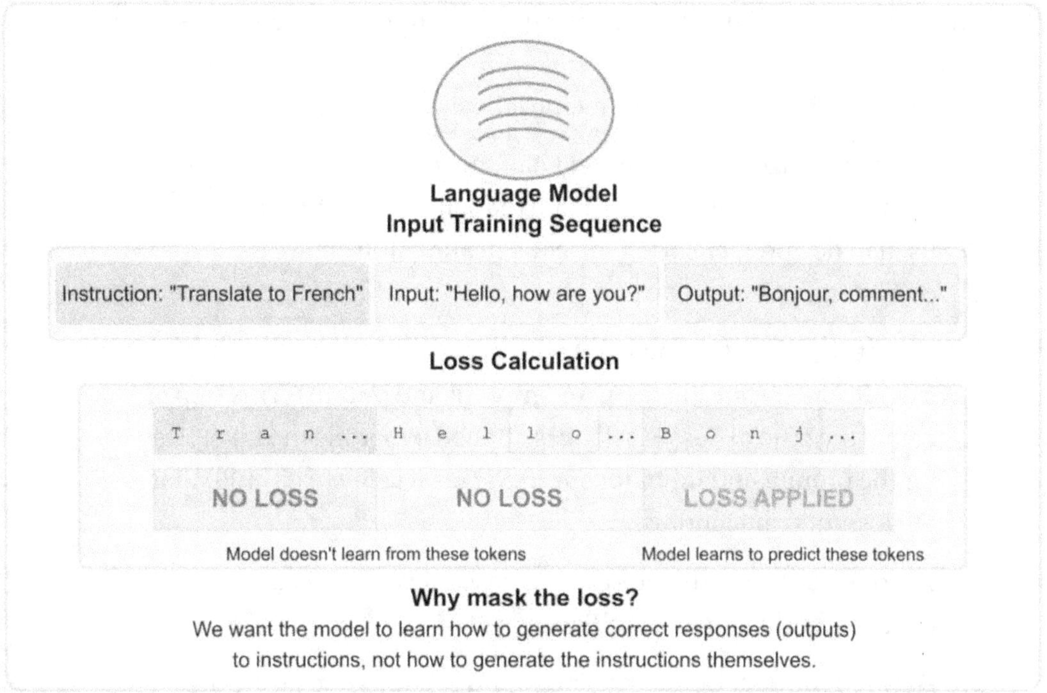

Figure 7-4. *Loss Masking in Instruction Fine-Tuning.This diagram illustrates how loss masking selectively trains language models during instruction fine-tuning. The model processes the entire instruction sequence, including input and output tokens, but the loss is calculated only on the output tokens (orange). No loss is applied to instruction or input tokens (pink/green), ensuring the model learns to generate appropriate responses rather than recreate the instructions themselves*

The following procedure demonstrates how to implement this training loop, which is the core of transforming a base model into an instruction-following assistant.

Recipe 2: The IFT Training Loop with Loss Masking

Fine-tuning a base Causal LM on a formatted instruction dataset using Trainer, incorporating loss masking.

Goal: Fine-tune a base Causal LM on a formatted instruction dataset using the Hugging Face Trainer, incorporating loss masking.

This procedure demonstrates the core of Instruction Fine-Tuning (IFT): implementing the training loop that transforms a base Causal Language Model (CLM) into an instruction-following assistant by predicting the next token while ignoring the prompt.

The fine-tuning workflow for this recipe follows these general steps:

1. **Load Tokenizer and Prepare:** Load the tokenizer for the base model (e.g., distilgpt2) and define the same Prompt Templates used in Recipe 1, ensuring the conversational structure is consistent.

2. **Create/Load and Format Dataset:** Tokenize the formatted text from Recipe 1 (the concatenated prompt and output). This step is where the crucial Loss Masking is applied by setting the label IDs for the prompt and input tokens to a special value (e.g., -100) that the loss function ignores.

3. **Load Model:** Load the pre-trained base Causal LM (e.g., distilgpt2) using AutoModelForCausalLM.

4. **Define Training:** Configure TrainingArguments (e.g., learning rate, batch size, and number of epochs) and initialize the Hugging Face Trainer with the model, tokenized dataset, and a standard Language Modeling data collator.

5. **Execute and Evaluate:** Execute the trainer.train() command to run the IFT process, followed by an optional evaluation (trainer. evaluate()) to report the validation loss.

What to Expect

This process culminates in an instruction-tuned model. By using loss masking, the model is trained to focus its learning solely on generating the tokens in the desired response (output) section. This mechanism explicitly teaches the base model to switch its behavior from simple text continuation to purposeful task completion.

```
# --- Recipe: The IFT Training Loop ---
# Goal: Fine-tune a base Causal LM on a formatted instruction dataset using
Trainer.
# Method: Includes prompt formatting and loss masking for
instruction tokens.
```

```python
# Libraries: transformers, datasets, torch, accelerate
# Note: Requires significant VRAM, especially for models like Gemma-2B.
#        Uses a tiny dummy dataset for demonstration. Replace with a real
           dataset.

from huggingface_hub import HfApi
from huggingface_hub import login

api = HfApi()
whoami = api.whoami(token="hf_xxxxxxxxxxxxxxxxxxxxxxx")
print(whoami)
login("hf_xxxxxxxxxxxxxxxxxxxxxxxxxx")

import torch
from datasets import Dataset # To create a dummy dataset
from transformers import (
    AutoTokenizer,
    AutoModelForCausalLM,
    TrainingArguments,
    Trainer,
    DataCollatorForLanguageModeling
)
import copy

# --- Configuration ---
MODEL_CHECKPOINT = "distilgpt2" # Use small model for demo; Replace with
e.g., "google/gemma-2b" for better results
# DATASET_NAME = "databricks/databricks-dolly-15k" # Example real dataset
(requires loading)
NUM_EPOCHS = 1
BATCH_SIZE = 2 # Keep very small for demo
GRADIENT_ACCUMULATION_STEPS = 2 # Simulate batch size 4
LEARNING_RATE = 2e-5
OUTPUT_DIR = "./instruction_finetune_output"
MAX_LENGTH = 256 # Max sequence length for tokenization
```

```python
# --- 1. Load Tokenizer & Prepare Prompt Templates (Alpaca Style) ---
print(f"Loading tokenizer for checkpoint: {MODEL_CHECKPOINT}")
try:
    tokenizer = AutoTokenizer.from_pretrained(MODEL_CHECKPOINT)
    if tokenizer.pad_token is None:
        tokenizer.pad_token = tokenizer.eos_token
        print(f"Set PAD token to EOS token: {tokenizer.pad_token}")
except Exception as e:
    print(f"Error loading tokenizer: {e}")
    exit()

PROMPT_WITH_INPUT_TEMPLATE = (
    "Below is an instruction that describes a task, paired with an input "
    "that provides further context. "
    "Write a response that appropriately completes the request.\n\n"
    "### Instruction:\n{instruction}\n\n### Input:\n{input}\n\n###"
    "Response:\n"
)
PROMPT_NO_INPUT_TEMPLATE = (
    "Below is an instruction that describes a task. "
    "Write a response that appropriately completes the request.\n\n"
    "### Instruction:\n{instruction}\n\n### Response:\n"
)

# --- 2. Create/Load and Format Dataset ---
# Using a tiny dummy dataset for demonstration
dummy_data = [
    {"instruction": "Say hello.", "input": "", "output": "Hello!"},
    {"instruction": "Add two numbers.", "input": "5 and 3", "output":
    "5 + 3 = 8"},
    {"instruction": "Describe the sun.", "input": "", "output": "The sun is
    a star at the center of our solar system."},
    {"instruction": "Translate to German.", "input": "Thank you", "output":
    "Danke schön"},
]
# Convert to Hugging Face Dataset object
```

```python
raw_dataset = Dataset.from_list(dummy_data)
print("Dummy Dataset:")
print(raw_dataset)

def format_and_tokenize(example):
    """Formats, tokenizes, and prepares labels with masking for IFT."""
    instruction = example.get("instruction", "")
    input_context = example.get("input", "")
    output = example.get("output", "")

    # Choose prompt template
    if input_context and input_context.strip():
        prompt_start = PROMPT_WITH_INPUT_TEMPLATE.
format(instruction=instruction, input=input_context)
    else:
        prompt_start = PROMPT_NO_INPUT_TEMPLATE.format(instruction=in
        struction)

    # Concatenate prompt and output, add EOS token
    full_text = prompt_start + output + tokenizer.eos_token

    # Tokenize the full text
    tokenized_full = tokenizer(full_text, truncation=True, padding="max_
    length", max_length=MAX_LENGTH)

   # Tokenize the prompt part *only* to find its true, unpadded length
   # CRITICAL: Use truncation=False and padding=False to get the true
     token count.
   tokenized_prompt = tokenizer(prompt_start, truncation=False,
   padding=False)
   prompt_length = len(tokenized_prompt["input_ids"])
   # Create labels - initially same as input_ids
   labels = copy.deepcopy(tokenized_full["input_ids"])

   # --- Crucial Step: Mask prompt tokens in labels ---
   # Set label IDs for prompt tokens to -100 so they are ignored in loss
     calculation
   # We use the unpadded prompt_length to mask the *padded* full_text.
```

```python
    for i in range(prompt_length):
        if i < len(labels):
            labels[i] = -100 # Check is technically not needed here, but
            is good practice

    # Ensure attention mask is included
    tokenized_full["labels"] = labels

    # Sanity check (optional): Decode labels, ignoring -100
    # decoded_labels = tokenizer.decode([l for l in labels if l != -100])
    # print(f"Decoded Labels (should match output + EOS): {decoded_
      labels}")

    return tokenized_full

print("\nFormatting and tokenizing dataset...")
try:
    # Apply formatting and tokenization
    # Remove original columns as they are now part of the tokenized
      structure
    tokenized_dataset = raw_dataset.map(
        format_and_tokenize,
        remove_columns=raw_dataset.column_names
    )
    tokenized_dataset.set_format("torch")
    # Split (if needed - using full small dataset for train/eval here
      for demo)
    # train_val_split = tokenized_dataset.train_test_split(test_size=0.1,
      seed=42)
    # final_datasets = {"train": train_val_split["train"], "validation":
      train_val_split["test"]}
    final_datasets = {"train": tokenized_dataset, "validation": tokenized_
    dataset} # Use same for demo

    print("Dataset processed. Sample tokenized input and labels:")
    print(f"Input IDs: {final_datasets['train'][0]['input_ids']}")
    print(f"Labels:    {final_datasets['train'][0]['labels']}") # Note
                        the -100 values
```

```python
except Exception as e:
    print(f"Error processing dataset: {e}")
    exit()

# --- 3. Data Collator ---
# Use standard LM collator; masking is handled in preprocessing
# mlm=False for Causal LM
data_collator = DataCollatorForLanguageModeling(tokenizer=tokenizer,
mlm=False)

# --- 4. Load Model ---
print(f"\nLoading base Causal LM: {MODEL_CHECKPOINT}")
try:
    model = AutoModelForCausalLM.from_pretrained(MODEL_CHECKPOINT)
    # Ensure model pad token id is set if tokenizer's was changed
    if model.config.pad_token_id is None:
        model.config.pad_token_id = tokenizer.pad_token_id
        print(f"Set model.config.pad_token_id to: {model.config.pad_
        token_id}")
    print(f"Model loaded. Initial device: {model.device}")
except Exception as e:
    print(f"Error loading model: {e}")
    exit()

# --- 5. Training Arguments ---
print("\nDefining Training Arguments...")
training_args = TrainingArguments(
    output_dir=OUTPUT_DIR,
    num_train_epochs=NUM_EPOCHS,
    per_device_train_batch_size=BATCH_SIZE,
    per_device_eval_batch_size=BATCH_SIZE,
    gradient_accumulation_steps=GRADIENT_ACCUMULATION_STEPS,
    learning_rate=LEARNING_RATE,
    weight_decay=0.01,
    eval_strategy="epoch",
    save_strategy="epoch",
    logging_strategy="steps",
```

```python
    logging_steps=1, # Log frequently for small dataset
    load_best_model_at_end=True,
    metric_for_best_model="loss",
    fp16=torch.cuda.is_available(),
    push_to_hub=False,
    report_to="none"
)
print(f"Training on device: {training_args.device}")

# --- 6. Initialize Trainer ---
trainer = Trainer(
    model=model,
    args=training_args,
    train_dataset=final_datasets["train"],
    eval_dataset=final_datasets["validation"],
    tokenizer=tokenizer,
    data_collator=data_collator,
    # No compute_metrics needed unless calculating perplexity explicitly
)

# --- 7. Train ---
print("\nStarting instruction fine-tuning...")
try:
    train_result = trainer.train()
    print("Training finished.")
    metrics = train_result.metrics
    trainer.log_metrics("train", metrics)
    trainer.save_metrics("train", metrics)
    trainer.save_state()
    trainer.save_model(OUTPUT_DIR) # Save final best model
    print(f"Model and training state saved to {OUTPUT_DIR}")
except Exception as e:
    print(f"Error during training: {e}")
    exit()
```

```python
# --- 8. Evaluate (Optional - reports loss by default) ---
print("\nEvaluating final model...")
try:
    eval_results = trainer.evaluate()
    print("Evaluation Results (Loss):")
    print(eval_results)
    # Calculate perplexity if desired
    # perplexity = math.exp(eval_results['eval_loss'])
    # print(f"Perplexity: {perplexity:.2f}")
except Exception as e:
    print(f"Error during evaluation: {e}")

# --- End of Recipe ---
```

Evaluating and Comparing IFT Models

After fine-tuning, it's crucial to evaluate the model's new capabilities. Unlike classification, where quantitative metrics like accuracy are straightforward, evaluating instruction-following often requires a **qualitative assessment**. We test the model on a diverse set of unseen prompts to assess whether it correctly interprets and executes instructions as intended. The following procedure shows how you would verify the results of the transformation performed in Recipe 2.

Recipe 3: Qualitative Evaluation of an IFT Model

Qualitatively evaluating the instruction-following capabilities of the fine-tuned model by testing it with various prompts. This procedure demonstrates how to verify the results of the Instruction Fine-Tuning (IFT) process performed in Recipe 2.

Goal: Test the instruction-following capabilities of the fine-tuned model with various unseen prompts to assess its new functionality.

The workflow for this recipe follows these general steps:

1. **Load Fine-Tuned Model and Tokenizer**: Load the saved IFT model and its corresponding tokenizer from the output directory (e.g., ./instruction_finetune_output).

2. **Create Inference Pipeline:** Utilize a simplified mechanism, such as the Hugging Face pipeline, to manage the generation process.

3. **Define Prompt Formatting Function (Crucial):** Re-implement the exact prompt templates (e.g., Alpaca style) used during the IFT training (Recipe 1).

4. **Test with Unseen Instructions:** Use a diverse set of prompts that the model did not see during training to assess its zero-shot performance and generalization.

5. **Qualitative Assessment:** Review the generated responses to determine if the model correctly interprets and executes the instructions, provides a relevant answer, and follows any formatting requests.

Critical Note on Inference Templating

It is essential to understand that the template used in IFT training (e.g., Alpaca, Vicuna, and ChatML) must be used during inference. If a user prompts an instruction-tuned model without the specific, structured template (e.g., by simply saying "Summarize this"), the model's performance will suffer dramatically. This is because the model won't recognize the specific start and end tokens, or the role separators, used in training to delineate the instruction from the expected response, causing it to fail to recognize and execute the task.

```
#Recipe: Qualitative Evaluation of IFT Model¶
# --- Recipe: Checking for Understanding (Qualitative Evaluation) ---
# Goal: Test the instruction-following capabilities of the fine-
tuned model.
# Method: Load the IFT model and generate responses for unseen
instructions.

import torch
from transformers import AutoTokenizer, AutoModelForCausalLM, pipeline

# --- Configuration ---
# Path to the saved instruction-fine-tuned model
IFT_MODEL_PATH = "./instruction_finetune_output" # From ch7_recipe_
ift_trainer
```

```python
# Base model checkpoint (needed to load tokenizer if not saved with model)
# Or load tokenizer from IFT_MODEL_PATH if it was saved there
TOKENIZER_CHECKPOINT = "distilgpt2" # Must match the base model
used for IFT

# Use GPU if available
device_index = 0 if torch.cuda.is_available() else -1

# --- 1. Load Fine-Tuned Model and Tokenizer ---
print(f"Loading fine-tuned IFT model from: {IFT_MODEL_PATH}")
try:
    model = AutoModelForCausalLM.from_pretrained(IFT_MODEL_PATH)
    # Try loading tokenizer from the fine-tuned path first,
      fallback to base
    try:
        tokenizer = AutoTokenizer.from_pretrained(IFT_MODEL_PATH)
    except OSError:
        print(f"Tokenizer not found in {IFT_MODEL_PATH}, loading from
        {TOKENIZER_CHECKPOINT}")
        tokenizer = AutoTokenizer.from_pretrained(TOKENIZER_CHECKPOINT)

    # Ensure PAD token is set correctly (important for generation)
    if tokenizer.pad_token is None:
        tokenizer.pad_token = tokenizer.eos_token
        print(f"Set PAD token to EOS token: {tokenizer.pad_token}")
    # Ensure model config has pad token id
    if model.config.pad_token_id is None:
        model.config.pad_token_id = tokenizer.pad_token_id

    print("IFT Model and Tokenizer loaded.")
except Exception as e:
    print(f"Error loading model/tokenizer: {e}")
    print("Ensure the IFT training recipe ran successfully and saved the
    model.")
    exit()
```

```python
# --- 2. Create Inference Pipeline ---
# Using pipeline simplifies generation
generator = pipeline(
    "text-generation",
    model=model,
    tokenizer=tokenizer,
    device=device_index
)

# --- 3. Define Prompt Formatting Function (Must match training template!) ---
# Re-use the templates from the formatting recipe
PROMPT_WITH_INPUT_TEMPLATE = (
    "Below is an instruction that describes a task, paired with an input "
    "that provides further context. "
    "Write a response that appropriately completes the request.\n\n"
    "### Instruction:\n{instruction}\n\n### Input:\n{input}\n\n###"
    "Response:\n"
)
PROMPT_NO_INPUT_TEMPLATE = (
    "Below is an instruction that describes a task. "
    "Write a response that appropriately completes the request.\n\n"
    "### Instruction:\n{instruction}\n\n### Response:\n"
)

def format_inference_prompt(instruction, input_context=""):
    """Formats the prompt for inference using the Alpaca template."""
    if input_context and input_context.strip():
        return PROMPT_WITH_INPUT_TEMPLATE.format(instruction=instruction,
        input=input_context)
    else:
        return PROMPT_NO_INPUT_TEMPLATE.format(instruction=instruction)

# --- 4. Test with Unseen Instructions ---
test_instructions = [
    {"instruction": "What is the capital of France?"},
    {"instruction": "Write a short story about a friendly robot.",
    "input": ""},
```

```python
    {"instruction": "Convert the following temperature from Celsius to
    Fahrenheit.", "input": "25°C"},
    {"instruction": "List the planets in our solar system."},
    {"instruction": "Generate a python function to calculate factorial"}
    # Task likely unseen in dummy data
]
print("\n--- Testing IFT Model with Unseen Instructions ---")
for i, test_case in enumerate(test_instructions):
    instruction = test_case["instruction"]
    input_context = test_case.get("input", "")

    # Format the prompt exactly as done during training (excluding
      the output)
    prompt = format_inference_prompt(instruction, input_context)

    print(f"\n--- Test Case {i+1} ---")
    print(f"Instruction: {instruction}")
    if input_context: print(f"Input: {input_context}")
    print(f"Formatted Prompt (Input to Model):\n{prompt}")
    print("-" * 20)

    try:
        # Generate response
        # Adjust generation parameters as needed
        outputs = generator(
            prompt,
            max_new_tokens=100, # Limit generated length
            num_return_sequences=1,
            do_sample=True,
            temperature=0.7,
            top_p=0.9,
            pad_token_id=tokenizer.eos_token_id # Often needed for
            Causal LMs
        )
```

```python
        # Extract only the generated part (the response)
        # The output includes the prompt, so we split/slice based on
          prompt length
        response = outputs[0]['generated_text'][len(prompt):].strip()

        print(f"Generated Response:\n{response}")

    except Exception as e:
        print(f"Error during generation for this test case: {e}")

    print("=" * 40)

# --- 5. Qualitative Assessment ---
# Review the generated responses. Does the model:
# - Understand the instruction?
# - Provide a relevant and coherent answer?
# - Follow formatting requests (if any were given)?
# - Avoid simply repeating the prompt?
# - Handle tasks it likely didn't see in the (dummy) training data?
#
# Note: With the dummy dataset and small model used in the training recipe,
# the results here will likely be poor. Using a larger base model
and a real
# instruction dataset (like Dolly) would yield much better instruction
following.

# --- End of Recipe ---
```

The source of the instruction dataset can also significantly impact the final model's behavior. A standard comparison is between datasets generated by other LLMs (**Self-Instruct**) and those created by humans (**Human-Curated**).

Recipe 4: Comparing Self-Instruct vs. Human-Curated Datasets

Fine-tuning the same base model on subsets of self-instruct vs. human-curated data and comparing qualitative results/validation loss. This recipe demonstrates how to assess the impact of the instruction dataset source on the final instruction-tuned model.

Goal: Fine-tune the same base model on two different dataset types (e.g., Alpaca vs. Dolly) and compare their performance qualitatively and via validation loss.

The fine-tuning workflow for this recipe follows these general steps:

1. **Load Tokenizer and Define Prompt Template:** Use the same tokenizer and prompt template (e.g., Alpaca-style) for both fine-tuning runs to ensure a consistent conversational structure and a fair comparison.

2. **Data Loading and Preprocessing Function:** Load subsets of both datasets (e.g., yahma/alpaca-cleaned and databricks/databricks-dolly-15k). The preprocessing function must adapt to the field names of each dataset and apply the same prompt formatting and loss masking to both.

3. **Fine-Tune on Dataset 1 (Self-Instruct):** Load the base model, define the training arguments (ensuring consistent parameters like learning rate and epochs), run the IFT trainer, and record the validation loss.

4. **Fine-Tune on Dataset 2 (Human-Curated):** Crucially, reload the BASE model before this step to prevent the training from continuing off the first model. Run the IFT trainer for the second dataset using the same training parameters and record its validation loss.

What to Expect

This process yields two distinct instruction-tuned models and a comparison of their evaluation metrics. The results provide a basis for a qualitative assessment of how dataset source—LLM-generated (Self-Instruct) vs. human-created (Human-Curated)—impacts the model's instruction-following capabilities, safety, and helpfulness, typically alongside a quantitative comparison of their validation loss.

Critical Note on Evaluation Robustness

For the most robust and unbiased comparison in a real-world or research setting, it is considered best practice to go beyond the simple train/validation split shown here. A more robust implementation would involve

1. Creating a third, completely separate Test Set that is entirely excluded from the training and validation splits of both datasets.

2. This test set should be saved (e.g., to a file) before running any fine-tuning process.

3. The final qualitative evaluation (and final quantitative performance report) must be performed only on this held-out Test Set to guarantee a clean, fair, and unbiased measure of each model's true zero-shot generalization capabilities.

```python
# --- Recipe: Comparing IFT Performance: Self-Instruct vs. Human-Curated
# Goal: Fine-tune the same base model on two dataset types (Alpaca-
style vs Dolly-style)and compare their performance qualitatively and via
validation loss.
# Method: Uses subsets of yahma/alpaca-cleaned and databricks/dolly-15k.
# Libraries: transformers, datasets, torch, accelerate
# Note: Uses distilgpt2 for speed; results differ greatly with
larger models.
# Requires sufficient disk space for datasets and model checkpoints.

import torch
import copy
import numpy as np
from datasets import load_dataset, Dataset
from transformers import (
    AutoTokenizer,
    AutoModelForCausalLM,
    TrainingArguments,
    Trainer,
    DataCollatorForLanguageModeling
)
import os
import random
```

```python
# --- Configuration ---
MODEL_CHECKPOINT = "distilgpt2" # Small model for faster demo
# Dataset 1: Self-Instruct style (Alpaca cleaned subset)
DATASET_1_NAME = "yahma/alpaca-cleaned"
OUTPUT_DIR_1 = "./ift_alpaca_output"
# Dataset 2: Human-Generated style (Dolly subset)
DATASET_2_NAME = "databricks/databricks-dolly-15k"
OUTPUT_DIR_2 = "./ift_dolly_output"

# Training Params (keep consistent for comparison)
NUM_EPOCHS = 1
BATCH_SIZE = 2 # Keep small for demo
GRADIENT_ACCUMULATION_STEPS = 4 # Effective batch size 8
LEARNING_RATE = 2e-5
MAX_LENGTH = 256 # Max sequence length
NUM_SAMPLES_PER_DATASET = 500 # Use small subsets for faster demo run

# --- 1. Load Tokenizer & Define Prompt Template ---
# Use the same tokenizer and template for both fine-tuning runs
print(f"Loading tokenizer for checkpoint: {MODEL_CHECKPOINT}")
try:
    tokenizer = AutoTokenizer.from_pretrained(MODEL_CHECKPOINT)
    if tokenizer.pad_token is None:
        tokenizer.pad_token = tokenizer.eos_token
        print(f"Set PAD token to EOS token: {tokenizer.pad_token}")
except Exception as e:
    print(f"Error loading tokenizer: {e}")
    exit()

# Using Alpaca-style template for consistency
PROMPT_WITH_INPUT_TEMPLATE = (
    "Below is an instruction that describes a task, paired with an input "
    "that provides further context. "
    "Write a response that appropriately completes the request.\n\n"
    "### Instruction:\n{instruction}\n\n### Input:\n{input}\n\n###"
    "Response:\n"
)
```

```python
PROMPT_NO_INPUT_TEMPLATE = (
    "Below is an instruction that describes a task. "
    "Write a response that appropriately completes the request.\n\n"
    "### Instruction:\n{instruction}\n\n### Response:\n"
)

# --- 2. Data Loading and Preprocessing Function ---
def format_and_tokenize(example, dataset_type):
    """Formats (Alpaca style), tokenizes, and prepares labels with
    masking."""
    # Adapt field names based on dataset
    if dataset_type == 'alpaca':
        instruction = example.get("instruction", "")
        input_context = example.get("input", "")
        output = example.get("output", "")
    elif dataset_type == 'dolly':
        instruction = example.get("instruction", "")
        input_context = example.get("context", "") # Dolly uses 'context'
        output = example.get("response", "")       # Dolly uses 'response'
    else:
        raise ValueError("Unknown dataset_type")

    # Choose prompt template
    if input_context and input_context.strip():
        prompt_start = PROMPT_WITH_INPUT_TEMPLATE.
        format(instruction=instruction, input=input_context)
    else:
        prompt_start = PROMPT_NO_INPUT_TEMPLATE.format(instruction=in
        struction)

    # Concatenate prompt and output, add EOS token
    full_text = prompt_start + output + tokenizer.eos_token

    # Tokenize the full text
    tokenized_full = tokenizer(full_text, truncation=True, padding="max_
    length", max_length=MAX_LENGTH)
```

```python
    # Tokenize the prompt part *only* to find its true, unpadded length
      for masking
    # CRITICAL: Use padding=False and truncation=False
    tokenized_prompt = tokenizer(prompt_start, truncation=False,
    padding=False)    prompt_length = len(tokenized_prompt["input_ids"])
     prompt_length = len(tokenized_prompt["input_ids"])

    # Create labels - initially same as input_ids
    labels = copy.deepcopy(tokenized_full["input_ids"])

    # --- Crucial Step: Mask prompt tokens in labels ---
    for i in range(prompt_length):
        if i < len(labels): # Ensure index is within bounds
            labels[i] = -100

    # Ensure attention mask is included
    tokenized_full["labels"] = labels
    return tokenized_full

# --- 3. Fine-Tune on Dataset 1 (Alpaca - Self-Instruct) ---
print(f"\n--- Processing Dataset 1: {DATASET_1_NAME} ---")
model_1_results = {}
try:
    # Load subset
    raw_dataset_1 = load_dataset(DATASET_1_NAME, split=f"train[:{NUM_
    SAMPLES_PER_DATASET}]")
    # Filter out examples that might be too long after formatting (optional
      but good practice)
    # raw_dataset_1 = raw_dataset_1.filter(lambda x: len(x['instruction'])
      + len(x['input']) + len(x['output']) < 1500)
    tokenized_dataset_1 = raw_dataset_1.map(
        lambda x: format_and_tokenize(x, 'alpaca'),
        remove_columns=raw_dataset_1.column_names
    )
    # Create dummy validation set for demo if needed
    split_ds_1 = tokenized_dataset_1.train_test_split(test_size=0.1,
    seed=42)
```

```python
final_datasets_1 = {"train": split_ds_1["train"], "validation": split_
ds_1["test"]}
print("Dataset 1 processed.")

# Load Base Model
model_1 = AutoModelForCausalLM.from_pretrained(MODEL_CHECKPOINT)
if model_1.config.pad_token_id is None: model_1.config.pad_token_id =
tokenizer.pad_token_id

# Training Args
training_args_1 = TrainingArguments(
    output_dir=OUTPUT_DIR_1, num_train_epochs=NUM_EPOCHS,
    per_device_train_batch_size=BATCH_SIZE, gradient_accumulation_
    steps=GRADIENT_ACCUMULATION_STEPS,
    learning_rate=LEARNING_RATE, weight_decay=0.01, eval_
    strategy="epoch",
    save_strategy="epoch", logging_strategy="steps", logging_steps=10,
    load_best_model_at_end=True, metric_for_best_model="loss",
    fp16=torch.cuda.is_available(), push_to_hub=False, report_to="none"
)

# Data Collator
data_collator = DataCollatorForLanguageModeling(tokenizer=tokenizer,
mlm=False)

# Trainer
trainer_1 = Trainer( model=model_1, args=training_args_1,
    train_dataset=final_datasets_1["train"], eval_dataset=final_
    datasets_1["validation"],
    tokenizer=tokenizer, data_collator=data_collator )

# Train
print(f"\nStarting fine-tuning on {DATASET_1_NAME}...")
train_result_1 = trainer_1.train()
trainer_1.save_model(OUTPUT_DIR_1)
print(f"Model 1 (Alpaca) saved to {OUTPUT_DIR_1}")
# Store eval results
```

```
eval_results_1 = trainer_1.evaluate(eval_dataset=final_
datasets_1["validation"])
model_1_results = {"eval_loss": eval_results_1.get("eval_loss", None)}
print(f"Model 1 (Alpaca) Validation Results: {model_1_results}")

except Exception as e:
    print(f"Error during Dataset 1 processing or training: {e}")
    # Clean up potentially loaded model
    if 'model_1' in locals(): del model_1
    torch.cuda.empty_cache()

# --- 4. Fine-Tune on Dataset 2 (Dolly - Human-Generated) ---
print(f"\n--- Processing Dataset 2: {DATASET_2_NAME} ---")
model_2_results = {}
# It's crucial to reload the *base* model to avoid continuing training
if os.path.exists(OUTPUT_DIR_2):
    print(f"Output directory {OUTPUT_DIR_2} already exists. Skipping
    training for Model 2 assuming it's done.")
    # Attempt to load previous results if needed for comparison later
    try:
        # Placeholder: In a real scenario you might load metrics saved
          by Trainer
        # For this demo, we'll assume it needs re-running if dir exists
          but no results loaded
        print("Cannot load previous results in this demo script. Re-run
        required if comparison needed.")
    except:
        print("Could not load previous results for Model 2.")

else:
    try:
        # Load subset
        raw_dataset_2 = load_dataset(DATASET_2_NAME, split=f"train[:{NUM_
        SAMPLES_PER_DATASET}]")
        # Filter out examples that might be too long after formatting
```

```python
# raw_dataset_2 = raw_dataset_2.filter(lambda x:
  len(x['instruction']) + len(x['context']) +
  len(x['response']) < 1500)
tokenized_dataset_2 = raw_dataset_2.map(
    lambda x: format_and_tokenize(x, 'dolly'),
    remove_columns=raw_dataset_2.column_names
)
split_ds_2 = tokenized_dataset_2.train_test_split(test_size=0.1,
seed=42)
final_datasets_2 = {"train": split_ds_2["train"], "validation":
split_ds_2["test"]}
print("Dataset 2 processed.")

# --- IMPORTANT: Reload the BASE model ---
print(f"Reloading BASE model: {MODEL_CHECKPOINT}")
model_2 = AutoModelForCausalLM.from_pretrained(MODEL_CHECKPOINT)
if model_2.config.pad_token_id is None: model_2.config.pad_token_id
= tokenizer.pad_token_id

# Training Args (use different output dir)
training_args_2 = TrainingArguments(
    output_dir=OUTPUT_DIR_2, num_train_epochs=NUM_EPOCHS,
    per_device_train_batch_size=BATCH_SIZE, gradient_accumulation_
    steps=GRADIENT_ACCUMULATION_STEPS,
    learning_rate=LEARNING_RATE, weight_decay=0.01, eval_
    strategy="epoch",
    save_strategy="epoch", logging_strategy="steps", logging_
    steps=10,
    load_best_model_at_end=True, metric_for_best_model="loss",
    fp16=torch.cuda.is_available(), push_to_hub=False, report_
    to="none"
)

# Data Collator (can reuse)
data_collator = DataCollatorForLanguageModeling(tokenizer=tokenizer,
mlm=False)
```

```python
        # Trainer
        trainer_2 = Trainer( model=model_2, args=training_args_2,
            train_dataset=final_datasets_2["train"], eval_dataset=final_
            datasets_2["validation"],
            tokenizer=tokenizer, data_collator=data_collator )

        # Train
        print(f"\nStarting fine-tuning on {DATASET_2_NAME}...")
        train_result_2 = trainer_2.train()
        trainer_2.save_model(OUTPUT_DIR_2)
        print(f"Model 2 (Dolly) saved to {OUTPUT_DIR_2}")
        # Store eval results
        eval_results_2 = trainer_2.evaluate(eval_dataset=final_
        datasets_2["validation"])
        model_2_results = {"eval_loss": eval_results_2.get("eval_
        loss", None)}
        print(f"Model 2 (Dolly) Validation Results: {model_2_results}")

    except Exception as e:
        print(f"Error during Dataset 2 processing or training: {e}")
        # Clean up
        if 'model_2' in locals(): del model_2
        torch.cuda.empty_cache()

# --- 5. Comparison ---
print("\n--- Comparison Summary ---")

# Compare Validation Loss (lower is generally better)
loss1 = model_1_results.get('eval_loss', 'N/A')
loss2 = model_2_results.get('eval_loss', 'N/A')
print(f"Model 1 (Alpaca) Final Validation Loss: {loss1}")
print(f"Model 2 (Dolly) Final Validation Loss: {loss2}")
if isinstance(loss1, float) and isinstance(loss2, float):
    if loss1 < loss2:
        print("Model 1 (Alpaca) had lower validation loss.")
    elif loss2 < loss1:
        print("Model 2 (Dolly) had lower validation loss.")
```

```python
    else:
        print("Validation losses were equal.")
else:
    print("Could not compare losses numerically.")

# Qualitative Evaluation on Sample Prompts
print("\n--- Qualitative Evaluation ---")
# Define a few diverse test prompts (ensure they weren't in the small
training subsets)
test_prompts = [
    {"instruction": "What are the primary colors?"},
    {"instruction": "Write a haiku about a cat."},
    {"instruction": "Explain the concept of recursion in programming."},
]

# Load models for inference if training was successful
model_inf_1 = None
model_inf_2 = None
if os.path.exists(OUTPUT_DIR_1):
    try:
        model_inf_1 = AutoModelForCausalLM.from_pretrained(OUTPUT_DIR_1)
        model_inf_1.to(training_args_1.device if 'training_args_1'
        in locals() else 'cpu') # Move to device
    except Exception as e: print(f"Failed to load Model 1: {e}")
if os.path.exists(OUTPUT_DIR_2):
    try:
        model_inf_2 = AutoModelForCausalLM.from_pretrained(OUTPUT_DIR_2)
        model_inf_2.to(training_args_2.device if 'training_args_2'
        in locals() else 'cpu') # Move to device
    except Exception as e: print(f"Failed to load Model 2: {e}")

# Helper function for generation
def generate_response(model, prompt_text):
    if model is None: return "Model not loaded."
    try:
        inputs = tokenizer(prompt_text, return_tensors="pt").
        to(model.device)
```

```python
        # Ensure pad token ID is set for generation
        gen_kwargs = {"max_new_tokens": 75, "pad_token_id": tokenizer.eos_
        token_id, "do_sample": True, "temperature": 0.7}
        outputs = model.generate(**inputs, **gen_kwargs)
        response = tokenizer.decode(outputs[0][inputs['input_ids'].
        shape[1]:], skip_special_tokens=True) # Decode only new tokens
        return response.strip()
    except Exception as e:
        return f"Generation Error: {e}"

# Format inference prompt (Alpaca style)
def format_inference_prompt(instruction, input_context=""):
    if input_context and input_context.strip():
        return PROMPT_WITH_INPUT_TEMPLATE.format(instruction=instruction,
        input=input_context)
    else:
        return PROMPT_NO_INPUT_TEMPLATE.format(instruction=instruction)
# Generate and compare
for i, p in enumerate(test_prompts):
    print(f"\n--- Test Prompt {i+1} ---")
    instruction = p['instruction']
    input_ctx = p.get('input', '')
    formatted_prompt = format_inference_prompt(instruction, input_ctx)
    print(f"Instruction: {instruction}")
    if input_ctx: print(f"Input: {input_ctx}")
    print("-" * 20)

    print("Model 1 (Alpaca) Response:")
    print(generate_response(model_inf_1, formatted_prompt))
    print("-" * 20)

    print("Model 2 (Dolly) Response:")
    print(generate_response(model_inf_2, formatted_prompt))
    print("=" * 40)
```

```
print("\nCompare the responses qualitatively: Which model seems more
helpful, accurate, creative, or better follows the instruction nuances?")
print("Note: Results heavily depend on base model size, dataset size/
quality, and training parameters.")

# --- End of Recipe ---
```

Chapter Summary

This chapter introduced you to Instruction Fine-Tuning (IFT), the crucial alignment process that transforms a raw, next-word-predicting base language model (like a Causal LM) into a responsive, helpful assistant. We established that this extra step is essential because it bridges the gap between a model's vast knowledge and its ability to act on explicit commands, dramatically improving its zero-shot instruction-following ability across diverse tasks. The core mechanism you learned involves training on a specialized, high-quality dataset containing instruction-input-output triplets and, critically, using loss masking to ensure the model's learning focuses solely on generating the appropriate response (the output) rather than just recreating the instruction. We also saw that IFT is the foundational first stage, followed by advanced alignment methods such as RLHF or DPO to refine the model's safety and helpfulness.

To successfully execute IFT, you must focus on the critical role of the dataset, its quality, diversity, and sourcing (whether through public, synthetic/Self-Instruct, or human-curated methods)—as this directly impacts the model's final behavior. We also demonstrated the necessity of consistent prompt templates (such as the Alpaca or Vicuna styles) for formatting your data, a structure that the model must recognize during both training and inference to correctly interpret and execute your commands. Now that you have explored both full fine-tuning and the general skill of instruction tuning, we will focus on making these powerful adaptation methods more accessible. The next chapter introduces Parameter-Efficient Fine-Tuning (PEFT), a set of techniques that significantly reduce the computational resources required to fine-tune large language models.

Parameter-Efficient Fine-Tuning (PEFT)

In our journey so far, we've witnessed the profound impact of full fine-tuning, a method that meticulously adapts every parameter of a model to a new task. This power, however, comes at a great computational cost, demanding significant GPU memory and time that can place the adaptation of large models out of reach for many. This chapter introduces **Parameter-Efficient Fine-Tuning (PEFT)**, a paradigm that encompasses a collection of brilliant techniques designed to make model adaptation more accessible and efficient.

The Rationale for PEFT: Adapting Models with Limited Resources

The primary challenge of full fine-tuning lies in its memory consumption. To update a model's weights, we must store not only the model itself but also the gradients and optimizer states for every single one of its billions of parameters. The PEFT paradigm offers an elegant solution. Instead of modifying all the weights, PEFT methods freeze the vast majority of the pre-trained base model's parameters and train only a very small number of new or adapted parameters—often less than 1% of the total.

This approach yields several profound advantages. It drastically reduces the memory footprint, making it feasible to fine-tune large models on consumer-grade GPUs. With fewer parameters to update, training is also significantly faster. Furthermore, since the original model weights are frozen, PEFT inherently mitigates the risk of "catastrophic forgetting." This also introduces a remarkable modularity; we can train small, lightweight "adapters" for different tasks and easily apply them to the same base model, which is far more efficient than storing multiple, massive fine-tuned models.

Low-Rank Adaptation (LoRA)

Currently, one of the most popular and successful PEFT methods is **Low-Rank Adaptation (LoRA)**. The key insight behind LoRA is the hypothesis that the weight changes needed to adapt a model for a new task have a "low intrinsic rank," meaning they can be effectively approximated by multiplying two much smaller, "thinner" matrices.

Instead of learning the massive weight change matrix directly, LoRA adds a parallel path to specific layers of the frozen base model (typically the attention layers). This path contains two small, trainable matrices (A and B). The final output of the layer becomes a combination of the original output and the output from this new LoRA path. Since only matrices A and B are trained, the number of trainable parameters is drastically reduced. The Figure 8-1 illustrates this concept by depicting how the input is processed by the large, frozen weight matrix W in parallel with the smaller trainable matrices A and B, whose outputs are combined to create the adapted result while training only a tiny fraction of the parameters.

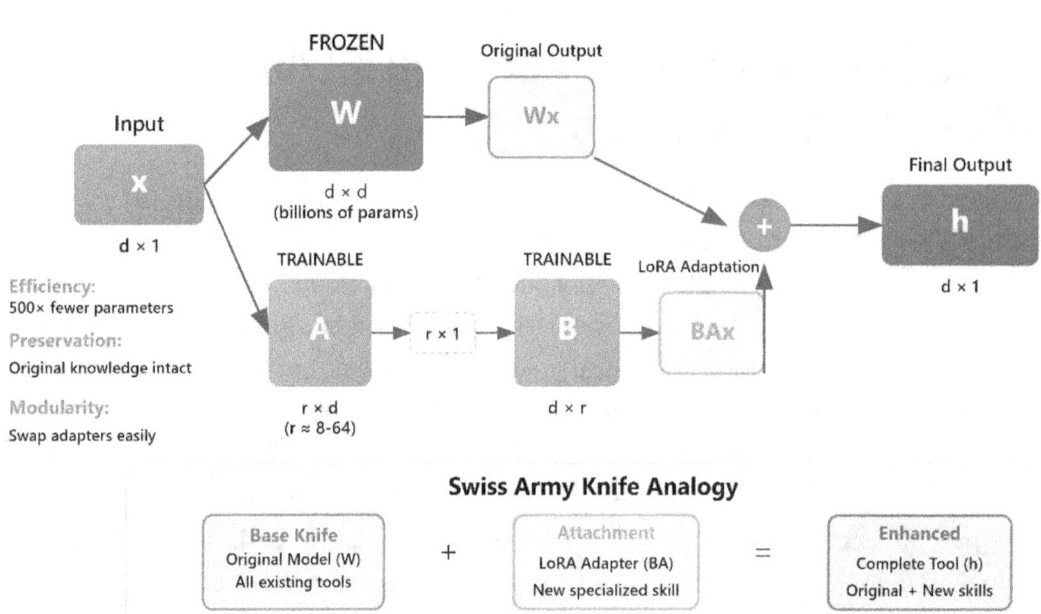

Figure 8-1. *LoRA: Low-Rank Adaptation Explained. This diagram shows how LoRA works. An input X is processed by the large, frozen weight matrix W. In parallel, it's processed by two small, trainable matrices A and B. The outputs are summed, yielding an adapted output while training only a tiny fraction of the parameters*

The core idea is simple:

A massive, pre-trained language model is like a brilliant student with a huge library of general knowledge.

- Full Fine-Tuning is like rewriting almost every book in the library on a single topic—it's exhaustive and slow.

- LoRA assumes that to adapt the model for a new task (like becoming a poetry expert), you don't need to rewrite the entire library.

- The "low intrinsic rank" hypothesis says the new knowledge for a specific task is just a small, focused "update note" or "steering correction." You only need two tiny new matrices (A and B) that act like a small adapter plugged into the main frozen model.

- This adapter provides the small, task-specific "steering signal" needed to fine-tune the model's behavior, which is much, much easier and faster than updating all the original, massive weights

Recipe 1: Standard LoRA Fine-Tuning

Fine-tuning a model using LoRA with the Hugging Face `peft` library

Goal: Fine-tune a base Causal Language Model (CLM) using **Standard Low-Rank Adaptation (LoRA)** on a formatted instruction dataset using the Hugging Face Trainer, incorporating loss masking.

This procedure demonstrates the core of **Standard Low-Rank Adaptation (LoRA) Fine-Tuning**: implementing the training loop that transforms a base Causal Language Model (CLM) into an instruction-following assistant by adapting only a minuscule fraction of its parameters.

The fine-tuning workflow for this recipe follows these general steps:

1. **Load Tokenizer and Prepare:** Load the tokenizer for the base model (e.g., `google/gemma-2b`) and define Prompt Templates to structure the instruction-response pairs (Alpaca/Dolly style).

2. **Create/Load and Format Dataset:** Tokenize the formatted text from the instruction dataset (e.g., Dolly subset). This step applies crucial **Loss Masking** by setting the label IDs for the prompt and input tokens to a special value (like -100), ensuring the loss calculation focuses solely on the desired response tokens.

3. **Load Base Model:** Load the pre-trained base Causal LM (e.g., Gemma-2B) using `AutoModelForCausalLM`, typically in a lower precision like bf16 or fp16 for memory efficiency

4. **Configure LoRA and Wrap Model:** Define the LoraConfig (specifying rank r, alpha, and target_modules) and then use get_peft_model to wrap the base model, freezing the original weights and introducing the small, trainable LoRA adapter matrices.

5. **Define Training and Initialize Trainer:** Configure `TrainingArguments` (e.g., learning rate, batch size, epochs, and gradient accumulation) and initialize the Hugging Face `Trainer` with the PEFT-wrapped model, tokenized dataset, and a standard Language Modeling data collator.

6. **Execute and Evaluate:** Execute the `trainer.train()` command to run the LoRA process, followed by an optional evaluation (`trainer.evaluate()`) to report the validation loss.

What to Expect

This process culminates in a parameter-efficient, instruction-tuned model. By using LoRA, the vast original knowledge of the base model is preserved (mitigating catastrophic forgetting), while the new, task-specific knowledge is captured in a lightweight "adapter." This allows the fine-tuned model to fit on more accessible hardware, and the only files saved are the small adapter weights, which are much more efficient than saving an entire copy of the massive fine-tuned model.

```
# --- Recipe: Standard LoRA Fine-Tuning ---
# Goal: Fine-tune a base model using LoRA with the Hugging Face PEFT
library.
# Method: Adapts the IFT recipe (Ch 7) using Gemma-2B and Dolly subset.
# Libraries: transformers, datasets, peft, torch, accelerate, bitsandbytes
(optional, for bf16/fp16)
# Note: Requires significant VRAM even without quantization for
Gemma-2B base.
#       Installs: pip install transformers datasets peft accelerate torch
        bitsandbytes sentencepiece

import torch
```

```python
import copy
from datasets import load_dataset, Dataset
from transformers import (
    AutoTokenizer,
    AutoModelForCausalLM,
    TrainingArguments,
    Trainer,
    DataCollatorForLanguageModeling,
    BitsAndBytesConfig # Used for loading in non-default dtypes like bf16
)
from peft import get_peft_model, LoraConfig, TaskType, prepare_model_for_
kbit_training

# --- Configuration ---
MODEL_CHECKPOINT = "google/gemma-2b" # Base model
DATASET_NAME = "databricks/databricks-dolly-15k" # Instruction dataset
OUTPUT_DIR = "./lora_finetune_output"
# LoRA Config
LORA_R = 16 # LoRA rank (e.g., 8, 16, 32, 64)
LORA_ALPHA = 32 # LoRA alpha (scaling factor, often 2*r)
LORA_DROPOUT =
.05
# Specify target modules for LoRA. Common practice for Gemma/Llama-
like models:
LORA_TARGET_MODULES = ["q_proj", "k_proj", "v_proj", "o_proj", "gate_proj",
"up_proj", "down_proj"]

# Training Params
NUM_EPOCHS = 1
BATCH_SIZE = 1 # Very small batch size for LoRA on Gemma-2B without QLoRA
GRADIENT_ACCUMULATION_STEPS = 16 # Effective batch size 1*16=16
LEARNING_RATE = 1e-4 # LoRA often uses higher LR than full FT
MAX_LENGTH = 512 # Max sequence length
NUM_SAMPLES_PER_DATASET = 500 # Use small subset for demo
```

```python
# --- 1. Load Tokenizer & Define Prompt Template (Same as IFT Chapter) ---
print(f"Loading tokenizer for checkpoint: {MODEL_CHECKPOINT}")
tokenizer = AutoTokenizer.from_pretrained(MODEL_CHECKPOINT)
if tokenizer.pad_token is None:
    tokenizer.pad_token = tokenizer.eos_token
    print(f"Set PAD token to EOS token: {tokenizer.pad_token}")

PROMPT_WITH_INPUT_TEMPLATE = (
    "Below is an instruction that describes a task, paired with an input "
    "that provides further context. "
    "Write a response that appropriately completes the request.\n\n"
    "### Instruction:\n{instruction}\n\n### Input:\n{input}\n\n###
Response:\n"
)
PROMPT_NO_INPUT_TEMPLATE = (
    "Below is an instruction that describes a task. "
    "Write a response that appropriately completes the request.\n\n"
    "### Instruction:\n{instruction}\n\n### Response:\n"
)

# --- 2. Data Loading and Preprocessing Function (Same as IFT Chapter) ---
def format_and_tokenize(example, dataset_type='dolly'): # Defaulting to
Dolly structure
    """Formats (Alpaca style), tokenizes, and prepares labels with masking
    for IFT."""
    instruction = example.get("instruction", "")
    input_context = example.get("context", "") # Dolly uses 'context'
    output = example.get("response", "")       # Dolly uses 'response'

    if input_context and input_context.strip():
        prompt_start = PROMPT_WITH_INPUT_TEMPLATE.
        format(instruction=instruction, input=input_context)
    else:
        prompt_start = PROMPT_NO_INPUT_TEMPLATE.format(instruction=instruction)
```

```
    full_text = prompt_start + output + tokenizer.eos_token
    tokenized_full = tokenizer(full_text, truncation=True, padding=False,
    max_length=MAX_LENGTH)
    tokenized_prompt = tokenizer(prompt_start, truncation=True,
    padding=False, max_length=MAX_LENGTH)
    prompt_length = len(tokenized_prompt["input_ids"])
    labels = copy.deepcopy(tokenized_full["input_ids"])

    # Mask prompt tokens
    for i in range(prompt_length):
        if i < len(labels): labels[i] = -100

    tokenized_full["labels"] = labels
    return tokenized_full

print(f"\n--- Processing Dataset: {DATASET_NAME} ---")
try:
    raw_dataset = load_dataset(DATASET_NAME, split=f"train[:{NUM_SAMPLES_
    PER_DATASET}]")
    tokenized_dataset = raw_dataset.map(
        format_and_tokenize, remove_columns=raw_dataset.column_names
    )
    split_ds = tokenized_dataset.train_test_split(test_size=0.1, seed=42)
    final_datasets = {"train": split_ds["train"], "validation": split_
    ds["test"]}
    print("Dataset processed.")
except Exception as e:
    print(f"Error processing dataset: {e}")
    exit()

# --- 3. Load Base Model (Not Quantized for standard LoRA) ---
print(f"\nLoading base model: {MODEL_CHECKPOINT}")
print("Loading in bf16/fp16 for memory efficiency...")
# Determine compute dtype
compute_dtype = torch.bfloat16 if torch.cuda.is_available() and torch.cuda.
is_bf16_supported() else torch.float16
```

```python
try:
    model = AutoModelForCausalLM.from_pretrained(
        MODEL_CHECKPOINT,
        torch_dtype=compute_dtype, # Load in lower precision
        device_map="auto" # Distribute across GPUs if available
    )
    # Ensure pad token ID is set
    if model.config.pad_token_id is None:
        model.config.pad_token_id = tokenizer.pad_token_id

    # Optional: Prepare model for k-bit training if using gradient
      checkpointing with lower precision loading
    # model = prepare_model_for_kbit_training(model, use_gradient_
      checkpointing=True) # Set use_gradient_checkpointing based on
      TrainingArgs

    print(f"Base model loaded in {compute_dtype}. Device map: {model.hf_
    device_map}")
except Exception as e:
    print(f"Error loading base model: {e}")
    print("Standard LoRA on Gemma-2B might still require > 24GB VRAM.
    Consider QLoRA.")
    exit()

# --- 4. Configure LoRA ---
print("\nConfiguring LoRA...")
peft_config = LoraConfig(
    r=LORA_R,
    lora_alpha=LORA_ALPHA,
    target_modules=LORA_TARGET_MODULES,
    lora_dropout=LORA_DROPOUT,
    bias="none", # Typically 'none', 'all', or 'lora_only'
    task_type=TaskType.CAUSAL_LM # Important for Causal LM tasks
)
```

```python
# --- 5. Wrap Model with PEFT ---
print("Applying PEFT LoRA adapters to the model...")
try:
    model = get_peft_model(model, peft_config)
    print("PEFT model created successfully.")
    model.print_trainable_parameters() # Shows the small percentage of
    trainable parameters
except Exception as e:
    print(f"Error applying PEFT: {e}")
    exit()

# --- 6. Training Arguments ---
print("\nDefining Training Arguments...")
training_args = TrainingArguments(
    output_dir=OUTPUT_DIR,
    num_train_epochs=NUM_EPOCHS,
    per_device_train_batch_size=BATCH_SIZE,
    gradient_accumulation_steps=GRADIENT_ACCUMULATION_STEPS,
    learning_rate=LEARNING_RATE,
    # Optimizer choice can matter, AdamW 8bit is memory efficient if
      bitsandbytes installed
    # optim="paged_adamw_8bit", # Or adamw_torch
    weight_decay=0.01,
    evaluation_strategy="epoch",
    save_strategy="epoch",
    logging_strategy="steps",
    logging_steps=10,
    load_best_model_at_end=True,
    metric_for_best_model="loss",
    fp16= (compute_dtype == torch.float16), # Enable based on compute_dtype
    bf16= (compute_dtype == torch.bfloat16), # Enable based on
    compute_dtype
    gradient_checkpointing=True, # Often needed for larger models/LoRA
    push_to_hub=False,
    report_to="none"
)
```

```python
# --- 7. Data Collator ---
data_collator = DataCollatorForLanguageModeling(tokenizer=tokenizer,
mlm=False)

# --- 8. Initialize Trainer ---
trainer = Trainer(
    model=model, # Pass the PEFT model
    args=training_args,
    train_dataset=final_datasets["train"],
    eval_dataset=final_datasets["validation"],
    tokenizer=tokenizer,
    data_collator=data_collator,
)

# --- 9. Train ---
print("\nStarting LoRA fine-tuning...")
try:
    train_result = trainer.train()
    # Note: PEFT saves only the adapter weights by default
    trainer.save_model(OUTPUT_DIR) # Saves adapter config & weights to
    OUTPUT_DIR
    print(f"LoRA training finished. Adapter saved to {OUTPUT_DIR}")
    # Log metrics
    metrics = train_result.metrics
    trainer.log_metrics("train", metrics)
    trainer.save_metrics("train", metrics)
    trainer.save_state()
except Exception as e:
    print(f"Error during LoRA training: {e}")
    exit()

# --- 10. Evaluate ---
print("\nEvaluating final LoRA model...")
try:
    eval_results = trainer.evaluate()
    print("Evaluation Results (Loss):")
    print(eval_results)
```

```
    # perplexity = math.exp(eval_results['eval_loss'])
    # print(f"Perplexity: {perplexity:.2f}")
    trainer.log_metrics("eval", eval_results)
    trainer.save_metrics("eval", eval_results)
except Exception as e:
    print(f"Error during evaluation: {e}")

# --- End of Recipe ---
```

QLoRA: Pushing Efficiency Further with Quantization

To push the boundaries of efficiency even further, the **QLoRA** technique combines the parameter-efficiency of LoRA with the memory-saving power of **quantization**. The process is ingenious: first, the large pre-trained base model is loaded with its weights quantized to a lower precision, typically 4-bit. This step alone dramatically reduces the memory required. This 4-bit model is then frozen, and standard LoRA adapters are added on top. During training, only the LoRA adapter weights (kept at a higher precision) are updated. This technique enables fine-tuning of very large models on consumer hardware. The Figure 8-2 depicts the QLoRA process, showing the large base model weights (W) loaded and stored in memory in 4-bit precision, with only the smaller, higher-precision LoRA adapters (BA) being updated during training.

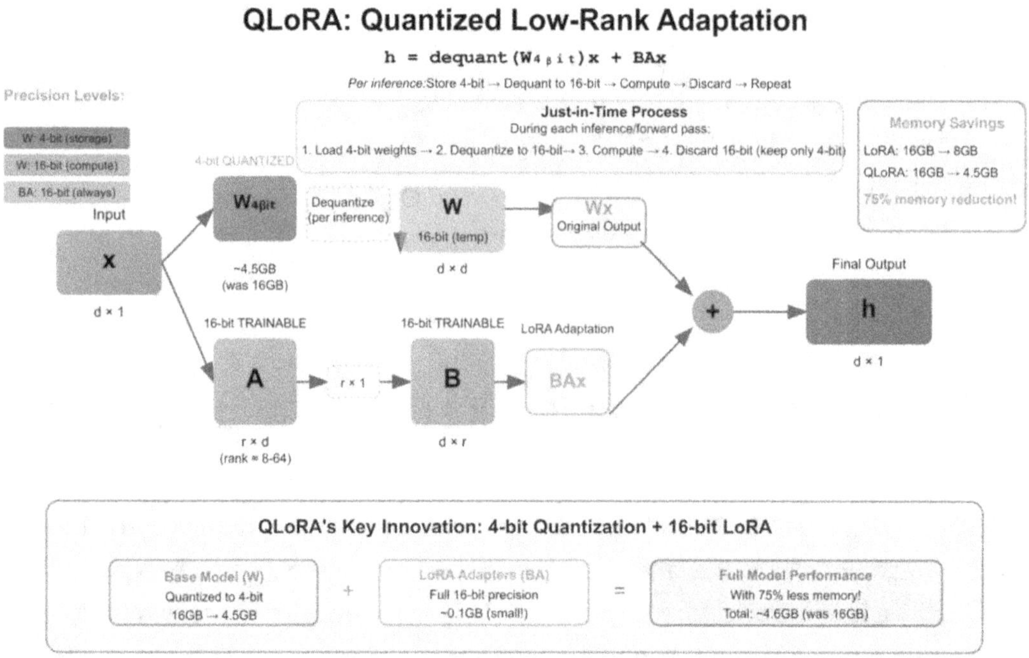

QLoRA: Quantized Low-Rank Adaptation
h = dequant(W₄ᵦᵢₜ)x + BAx
Per inference:Store 4-bit → Dequant to 16-bit → Compute → Discard → Repeat
Precision Levels:
W: 4-bit (storage)
W: 16-bit (compute)
BA: 16-bit (always)
Input
Just-in-Time Process
During each inference/forward pass:
1. Load 4-bit weights → 2. Dequantize to 16-bit → 3. Compute → 4. Discard 16-bit (keep only 4-bit)
Memory Savings
LoRA: 16GB → 8GB
QLoRA: 16GB → 4.5GB
75% memory reduction!
4-bit QUANTIZED
W₄ᵦᵢₜ
Dequantize
(per inference)
W
16-bit (temp)
Wx
Original Output
x
d × 1
~4.5GB
(was 16GB)
d × d
Final Output
16-bit TRAINABLE
A
r × 1
16-bit TRAINABLE
B
LoRA Adaptation
BAx
+
h
d × 1
r × d
(rank ≈ 8-64)
d × r
QLoRA's Key Innovation: 4-bit Quantization + 16-bit LoRA
Base Model (W)
Quantized to 4-bit
16GB → 4.5GB
+
LoRA Adapters (BA)
Full 16-bit precision
~0.1GB (small!)
=
Full Model Performance
With 75% less memory!
Total: ~4.6GB (was 16GB)

***Figure 8-2.** QLoRA: Quantized Low-Rank Adaptation. This diagram illustrates the QLoRA process. The large base model (W) is loaded and stored in 4-bit precision. During the forward pass, these weights are de-quantized just in time for computation and then discarded, while the small LoRA adapters (BA) remain in higher precision and are the only parameters trained*

Recipe 2: QLoRA on a Budget

Fine-tuning using QLoRA (4-bit quantization + LoRA) with `bitsandbytes` and `peft`.

Goal: Fine-tune a base model using **QLoRA** (4-bit quantization + LoRA) with the bitsandbytes and peft libraries.

This procedure demonstrates the core of **Quantized Low-Rank Adaptation (QLoRA):** implementing the training loop that combines LoRA's parameter efficiency with 4-bit quantization, enabling the fine-tuning of very large models on memory-constrained consumer hardware.

The fine-tuning workflow for this recipe follows these general steps:

1. **Load Tokenizer and Prepare:** Load the tokenizer for the base model (e.g., `google/gemma-2b`) and define Prompt Templates for instruction-response formatting.

2. **Create/Load and Format Dataset:** Tokenize the instruction dataset (e.g., Dolly subset), applying **Loss Masking** to ensure the loss calculation focuses solely on the desired response tokens.

3. **Configure Quantization:** Define the `BitsAndBytesConfig`, setting the base model to be loaded in 4-bit precision (e.g., `load_in_4bit=True`).

4. **Load Base Model in 4-bit:** Load the pre-trained base Causal LM (e.g., Gemma-2B) using `AutoModelForCausalLM` and the defined quantization config.

5. **Prepare Model for PEFT:** Call `prepare_model_for_kbit_training` and enable gradient checkpointing for maximal memory savings.

6. **Configure LoRA and Wrap Model:** Define the `LoraConfig` (specifying rank r, alpha, and target_modules) and then use `get_peft_model` to wrap the 4-bit base model, introducing the small, high-precision LoRA adapter matrices.

7. **Define Training and Initialize Trainer:** Configure `TrainingArguments` (often using a paged optimizer like `paged_adamw_8bit` for extra memory efficiency) and initialize the Hugging Face `Trainer` with the QLoRA-wrapped model.

8. **Execute and Evaluate:** Execute the `trainer.train()` command to run the QLoRA process, followed by an optional evaluation (`trainer.evaluate()`).

What to Expect

This process culminates in a parameter-efficient, instruction-tuned model that requires dramatically less VRAM (typically 12–16GB or less) compared to standard LoRA or Full Fine-Tuning. The base model's knowledge is preserved, and the new, task-specific knowledge is captured in the lightweight, higher-precision LoRA adapter weights.

```
# --- Recipe: QLoRA on a Budget ---
# Goal: Fine-tune a base model using QLoRA (4-bit quantization + LoRA).
# Method: Uses Gemma-2B, Dolly subset, bitsandbytes for 4-bit loading,
and peft.
```

```python
# Libraries: transformers, datasets, peft, accelerate, torch, bitsandbytes,
    sentencepiece
# Note: Should fit in GPUs with >= 12-16GB VRAM. Install bitsandbytes: pip
    install bitsandbytes

import torch
import copy
from datasets import load_dataset, Dataset
from transformers import (
    AutoTokenizer,
    AutoModelForCausalLM,
    TrainingArguments,
    Trainer,
    DataCollatorForLanguageModeling,
    BitsAndBytesConfig # Needed for quantization config
)
from peft import get_peft_model, LoraConfig, TaskType, prepare_model_for_
kbit_training

# --- Configuration (Similar to LoRA recipe) ---
MODEL_CHECKPOINT = "google/gemma-2b"
DATASET_NAME = "databricks/databricks-dolly-15k"
OUTPUT_DIR = "./qlora_finetune_output"
# LoRA Config
LORA_R = 16
LORA_ALPHA = 32
LORA_DROPOUT = 0.05
LORA_TARGET_MODULES = ["q_proj", "k_proj", "v_proj", "o_proj", "gate_proj",
"up_proj", "down_proj"]
# Training Params
NUM_EPOCHS = 1
# Can potentially use larger batch size with QLoRA vs standard LoRA
BATCH_SIZE = 2 # Start small, increase if memory allows
GRADIENT_ACCUMULATION_STEPS = 8 # Effective batch size 2*8=16
LEARNING_RATE = 1e-4 # QLoRA might tolerate slightly higher LR sometimes
MAX_LENGTH = 512
```

```python
NUM_SAMPLES_PER_DATASET = 500

# --- 1. Load Tokenizer & Define Prompt Template (Same as LoRA recipe) ---
print(f"Loading tokenizer for checkpoint: {MODEL_CHECKPOINT}")
tokenizer = AutoTokenizer.from_pretrained(MODEL_CHECKPOINT)
if tokenizer.pad_token is None:
    tokenizer.pad_token = tokenizer.eos_token

PROMPT_WITH_INPUT_TEMPLATE = (
    "Below is an instruction that describes a task, paired with an input"
    "that provides further context. "
    "Write a response that appropriately completes the request.\n\n"
    "### Instruction:\n{instruction}\n\n### Input:\n{input}\n\n###"
    "Response:\n"
)
PROMPT_NO_INPUT_TEMPLATE = (
    "Below is an instruction that describes a task. "
    "Write a response that appropriately completes the request.\n\n"
    "### Instruction:\n{instruction}\n\n### Response:\n"
)

# --- 2. Data Loading and Preprocessing Function (Same as LoRA recipe) ---
def format_and_tokenize(example, dataset_type='dolly'):
    instruction = example.get("instruction", "")
    input_context = example.get("context", "")
    output = example.get("response", "")
    if input_context and input_context.strip():
        prompt_start = PROMPT_WITH_INPUT_TEMPLATE.
        format(instruction=instruction, input=input_context)
    else:
        prompt_start = PROMPT_NO_INPUT_TEMPLATE.format(instruction=in
        struction)
    full_text = prompt_start + output + tokenizer.eos_token
    tokenized_full = tokenizer(full_text, truncation=True, padding=False,
    max_length=MAX_LENGTH)
    tokenized_prompt = tokenizer(prompt_start, truncation=True,
    padding=False, max_length=MAX_LENGTH)
```

```python
    prompt_length = len(tokenized_prompt["input_ids"])
    labels = copy.deepcopy(tokenized_full["input_ids"])
    for i in range(prompt_length):
        if i < len(labels): labels[i] = -100
    tokenized_full["labels"] = labels
    return tokenized_full

print(f"\n--- Processing Dataset: {DATASET_NAME} ---")
try:
    raw_dataset = load_dataset(DATASET_NAME, split=f"train[:{NUM_SAMPLES_
    PER_DATASET}]")
    tokenized_dataset = raw_dataset.map(
        format_and_tokenize, remove_columns=raw_dataset.column_names
    )
    split_ds = tokenized_dataset.train_test_split(test_size=0.1, seed=42)
    final_datasets = {"train": split_ds["train"], "validation": split_
    ds["test"]}
    print("Dataset processed.")
except Exception as e:
    print(f"Error processing dataset: {e}")
    exit()

# --- 3. Configure Quantization (BitsAndBytes) ---
print("\nConfiguring 4-bit quantization...")
# Determine compute dtype
compute_dtype = torch.bfloat16 if torch.cuda.is_available() and torch.cuda.
is_bf16_supported() else torch.float16
print(f"Using compute dtype: {compute_dtype}")

bnb_config = BitsAndBytesConfig(
    load_in_4bit=True,
    bnb_4bit_quant_type="nf4", # Use NF4 (NormalFloat4) data type for
    optimal results
    bnb_4bit_compute_dtype=compute_dtype, # Computations done in bf16/fp16
    bnb_4bit_use_double_quant=True, # Optional: Use double quantization for
    extra memory savings
)
```

```python
# --- 4. Load Base Model in 4-bit ---
print(f"\nLoading base model ({MODEL_CHECKPOINT}) in 4-bit...")
try:
    model = AutoModelForCausalLM.from_pretrained(
        MODEL_CHECKPOINT,
        quantization_config=bnb_config, # Apply quantization config
        device_map="auto" # Automatically distribute quantized model
        # trust_remote_code=True # Add if needed for specific models
          like Phi
    )
    # Ensure pad token ID is set
    if model.config.pad_token_id is None:
        model.config.pad_token_id = tokenizer.pad_token_id

    print(f"Base model loaded in 4-bit. Device map: {model.hf_device_map}")
except Exception as e:
    print(f"Error loading base model in 4-bit: {e}")
    print("Ensure 'bitsandbytes' is installed correctly for your CUDA
    version.")
    exit()

# --- 5. Prepare Model for PEFT & Configure LoRA ---
# Prepare for k-bit training must be called BEFORE applying PEFT config
# Enable gradient checkpointing for more memory savings
model = prepare_model_for_kbit_training(model, use_gradient_
checkpointing=True)
print("Model prepared for k-bit training.")

print("\nConfiguring LoRA...")
peft_config = LoraConfig(
    r=LORA_R,
    lora_alpha=LORA_ALPHA,
    target_modules=LORA_TARGET_MODULES, # Target modules might differ
    slightly per model, check docs
    lora_dropout=LORA_DROPOUT,
    bias="none",
    task_type=TaskType.CAUSAL_LM
 )
```

```python
# --- 6. Wrap Model with PEFT ---
print("Applying PEFT LoRA adapters to the 4-bit model...")
try:
    model = get_peft_model(model, peft_config)
    print("PEFT model created successfully.")
    model.print_trainable_parameters() # Shows the small percentage of
    trainable parameters
except Exception as e:
    print(f"Error applying PEFT: {e}")
    exit()

# --- 7. Training Arguments ---
# Use paged optimizer for more memory efficiency with QLoRA
use_paged_optimizer = True # Requires bitsandbytes >= 0.41.1
optim_choice = "paged_adamw_8bit" if use_paged_optimizer else "adamw_torch"
print(f"Using optimizer: {optim_choice}")

print("\nDefining Training Arguments...")
training_args = TrainingArguments(
    output_dir=OUTPUT_DIR,
    num_train_epochs=NUM_EPOCHS,
    per_device_train_batch_size=BATCH_SIZE,
    gradient_accumulation_steps=GRADIENT_ACCUMULATION_STEPS,
    learning_rate=LEARNING_RATE,
    optim=optim_choice, # Use paged optimizer
    weight_decay=0.01,
    evaluation_strategy="epoch",
    save_strategy="epoch",
    logging_strategy="steps",
    logging_steps=10,
    load_best_model_at_end=True,
    metric_for_best_model="loss",
    fp16= (compute_dtype == torch.float16), # Enable based on compute_dtype
    bf16= (compute_dtype == torch.bfloat16), # Enable based on
    compute_dtype
    gradient_checkpointing=True, # Crucial for QLoRA memory saving
```

```python
    push_to_hub=False,
    report_to="none"
)

# --- 8. Data Collator ---
data_collator = DataCollatorForLanguageModeling(tokenizer=tokenizer,
mlm=False)

# --- 9. Initialize Trainer ---
trainer = Trainer(
    model=model, # Pass the QLoRA PEFT model
    args=training_args,
    train_dataset=final_datasets["train"],
    eval_dataset=final_datasets["validation"],
    tokenizer=tokenizer,
    data_collator=data_collator,
)

# --- 10. Train ---
print("\nStarting QLoRA fine-tuning...")
try:
    train_result = trainer.train()
    trainer.save_model(OUTPUT_DIR) # Saves adapter config & weights
    print(f"QLoRA training finished. Adapter saved to {OUTPUT_DIR}")
    metrics = train_result.metrics
    trainer.log_metrics("train", metrics)
    trainer.save_metrics("train", metrics)
    trainer.save_state()
except Exception as e:
    print(f"Error during QLoRA training: {e}")
    exit()

# --- 11. Evaluate ---
print("\nEvaluating final QLoRA model...")
try:
    eval_results = trainer.evaluate()
    print("Evaluation Results (Loss):")
```

```
    print(eval_results)
    trainer.log_metrics("eval", eval_results)
    trainer.save_metrics("eval", eval_results)
except Exception as e:
    print(f"Error during evaluation: {e}")

# --- End of Recipe ---
```

The contrast between narrowly targeting the attention mechanisms and adopting a full-coverage strategy for LoRA fine-tuning is a classic trade-off between parsimony and potency.

Targeting only the query (q_proj) and value (v_proj) projection matrices in a model's attention layers for LoRA fine-tuning is a common strategy that represents a higher efficiency trade-off. The rationale is that these matrices are most critical for shaping a model's understanding of new, task-specific input, thereby maximizing the impact of the small parameter update within a given computational budget. The benefit is the minimal number of trainable parameters and the fastest training speed. Conversely, adopting "full coverage," which includes targeting all attention layers (e.g., k_proj for keys and o_proj for outputs) and the essential feed-forward network matrices (gate_proj, up_proj, down_proj), introduces a trade-off between higher model quality and more thorough adaptation (used in this recipe). While this approach increases the number of trainable parameters (though still less than 1% of the base model's weights) and requires slightly more VRAM and time, it typically yields a small but measurable performance gain by allowing the adapter to more fully modulate the behavior of the entire Transformer block. The choice depends on hardware constraints and the performance gap of the specific task; for high-stakes tasks, full coverage is often preferred, while efficiency-first applications often stick to q_proj and v_proj.

Other PEFT Methods

While LoRA is dominant, other methods are also part of the PEFT family. **Adapters**, one of the earlier techniques, involve inserting small, new feed-forward layers sequentially within the Transformer blocks while keeping the base model frozen. **Prefix Tuning** takes a different approach entirely; instead of modifying any weights, it trains a small set of continuous vector embeddings (a "soft prompt") that are prepended to the input sequence, effectively steering the frozen model's behavior. However, we are not going to discuss these techniques deeply.

While these core techniques are powerful on their own, the quest for even greater speed and efficiency has led to further innovations in the form of specialized acceleration libraries.

Accelerating PEFT with UnslothAI

While QLoRA is highly efficient, specialized libraries can further improve performance. **UnslothAI** is an open source library designed specifically to accelerate LoRA and QLoRA fine-tuning, often delivering substantial improvements in both speed and memory efficiency. It achieves this through highly optimized, hand-written computational kernels for key operations. By replacing generic PyTorch operations with these custom alternatives, UnslothAI can significantly reduce training time and memory consumption, often allowing for larger batch sizes on the same hardware with minimal code changes. The Figure 8-3 shows how libraries like UnslothAI accelerate PEFT by utilizing highly optimized computational kernels to replace generic PyTorch operations, resulting in significant speed and memory efficiency improvements during fine-tuning.

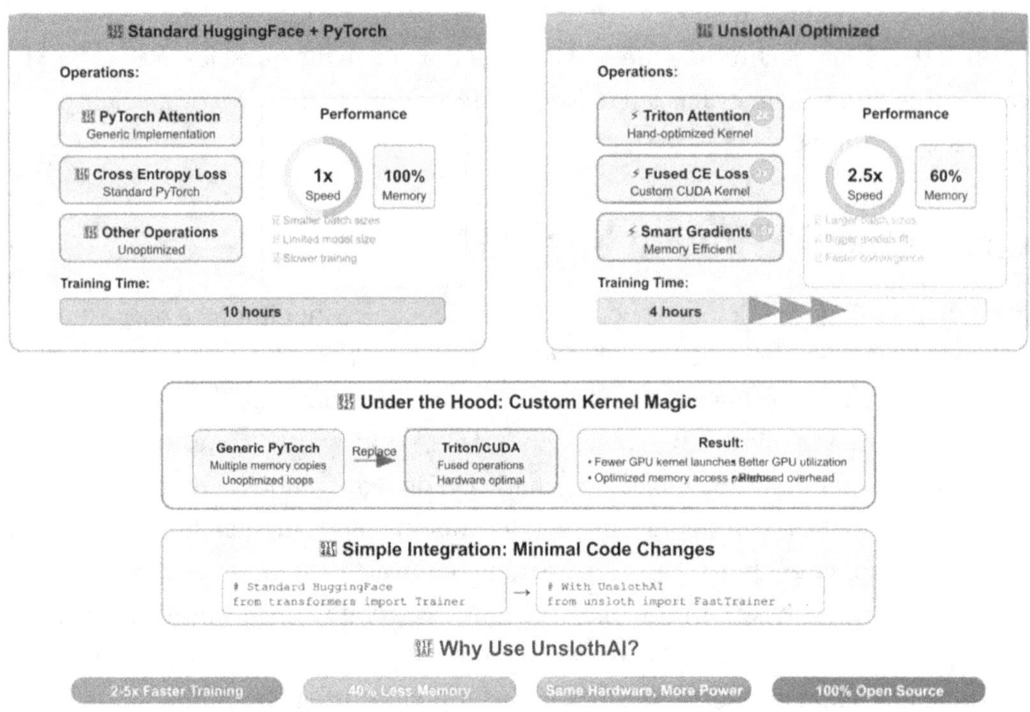

Figure 8-3. *Comparing and Contrasting UnslothAI vs. Standard Hugging Face + PyTorch*

The technical optimizations in UnslothAI focus on replacing high-level, generic operations with low-level, hardware-aware kernels. Using Triton and custom CUDA kernels, UnslothAI implements Fused Cross-Entropy Loss and Triton-optimized Attention, which combine multiple mathematical steps into a single GPU operation. This "fusion" is relevant because it minimizes expensive memory read/writes (VRAM bottlenecks) and reduces the overhead of launching thousands of individual GPU kernels. Additionally, Smart Gradient management optimizes backpropagation, leading to a 40% reduction in memory footprint and 2.5x faster training. For developers, these optimizations are significant because they enable training larger models or using larger batch sizes on the same consumer-grade hardware, effectively lowering the barrier to entry for high-performance LLM fine-tuning.

Recipe 3: Accelerated PEFT with UnslothAI

Speeding up QLoRA fine-tuning using the UnslothAI library

Goal: Demonstrate accelerated QLoRA fine-tuning using the UnslothAI library.

This procedure demonstrates the core of **Accelerated PEFT with UnslothAI**: implementing a turbocharged fine-tuning loop that leverages UnslothAI's specialized, optimized kernels to dramatically increase the speed and memory efficiency of the QLoRA process.

The fine-tuning workflow for this recipe follows these general steps:

1. **Load Tokenizer and Prepare:** Load the tokenizer for the base model (e.g., `google/gemma-2b`) and define Prompt Templates for instruction-response formatting.

2. **Create/Load and Format Dataset:** Tokenize the instruction dataset (e.g., Dolly subset), applying **Loss Masking** to ensure the loss calculation focuses solely on the desired response tokens.

3. **Load Model with UnslothAI:** Load the base model using the UnslothAI function `FastLanguageModel.from_pretrained`, which internally handles the 4-bit quantization and initial optimizations.

4. **Apply PEFT (LoRA) Configuration:** Apply the LoRA configuration (specifying rank r, alpha, and target_modules) using the optimized UnslothAI function `FastLanguageModel.get_peft_model` to integrate the LoRA adapters into the 4-bit base model.

5. **Define Training and Initialize Trainer:** Configure `TrainingArguments` (using `paged_adamw_8bit` and Unsloth's built-in bfloat16/fp16 checks) and initialize the standard Hugging Face `Trainer`.

6. **Execute and Evaluate:** Execute the `trainer.train()` command, followed by an optional evaluation (`trainer.evaluate()`).

What to Expect

This process culminates in the fastest and most memory-efficient fine-tuned model possible for the specified hardware constraints. The key benefit of UnslothAI's specialized kernels is a significant acceleration (up to 2.5x faster training) and a further reduction in memory footprint (up to 40% less VRAM) compared to standard QLoRA. This allows you to train larger models or use larger batch sizes on consumer-grade GPUs, with the resulting adapter saved in a standard, compatible PEFT format.

```
# --- Recipe: Turbocharged PEFT with UnslothAI ---
# Goal: Demonstrate accelerated LoRA/QLoRA fine-tuning using UnslothAI.
# Method: Adapts the QLoRA recipe showing minimal code changes for Unsloth.
# Libraries: unsloth, torch, datasets, peft, transformers, sentencepiece
# Note: Requires UnslothAI installation specific to your environment (GPU/
  PyTorch/CUDA).
#         Example install: pip install "unsloth[pytorch-ampere-torch210]"
          --extra-index-url https://pypi.unsloth.ai
#         Check UnslothAI GitHub for correct install command.

import torch
import copy
from datasets import load_dataset, Dataset
from transformers import (
    AutoTokenizer,
    TrainingArguments,
    Trainer,
    DataCollatorForLanguageModeling,
    BitsAndBytesConfig # Still used for quantization config if desired
)
# --- UnslothAI Import ---
from unsloth import FastLanguageModel # Main import
from unsloth import is_bfloat16_supported
from peft import LoraConfig, TaskType # PEFT imports remain similar

# --- Configuration (Similar to QLoRA recipe) ---
MODEL_CHECKPOINT = "google/gemma-2b" # Unsloth supports many models
DATASET_NAME = "databricks/databricks-dolly-15k"
OUTPUT_DIR = "./unsloth_qlora_finetune_output"
# LoRA Config (Same as before)
LORA_R = 16
LORA_ALPHA = 32
LORA_DROPOUT = 0.05
```

```python
# Target modules might be automatically inferred by Unsloth for
  some models,
# but specifying is often still good practice or required. Check
  Unsloth docs.
LORA_TARGET_MODULES = ["q_proj", "k_proj", "v_proj", "o_proj", "gate_proj",
"up_proj", "down_proj"]
# Training Params
NUM_EPOCHS = 1
# Unsloth's memory savings might allow larger batch sizes vs standard QLoRA
BATCH_SIZE = 4 # Try increasing this (e.g., 8, 16) if memory allows!
GRADIENT_ACCUMULATION_STEPS = 4 # Adjust so BATCH_SIZE * GRAD_ACCUM is your
target effective batch size
LEARNING_RATE = 1e-4
MAX_LENGTH = 512
NUM_SAMPLES_PER_DATASET = 500


# --- 1. Load Tokenizer & Define Prompt Template (Same as LoRA/QLoRA recipe) ---
print(f"Loading tokenizer for checkpoint: {MODEL_CHECKPOINT}")
tokenizer = AutoTokenizer.from_pretrained(MODEL_CHECKPOINT)
if tokenizer.pad_token is None:
    tokenizer.pad_token = tokenizer.eos_token

PROMPT_WITH_INPUT_TEMPLATE = (
    "Below is an instruction that describes a task, paired with an input
    that provides further context. "
    "Write a response that appropriately completes the request.\n\n"
    "### Instruction:\n{instruction}\n\n### Input:\n{input}\n\n###
    Response:\n"
)
PROMPT_NO_INPUT_TEMPLATE = (
    "Below is an instruction that describes a task. "
    "Write a response that appropriately completes the request.\n\n"
    "### Instruction:\n{instruction}\n\n### Response:\n"
)
```

```python
# --- 2. Data Loading and Preprocessing Function (Same as LoRA/QLoRA recipe) ---
def format_and_tokenize(example, dataset_type='dolly'):
    instruction = example.get("instruction", "")
    input_context = example.get("context", "")
    output = example.get("response", "")
    if input_context and input_context.strip():
        prompt_start = PROMPT_WITH_INPUT_TEMPLATE.
        format(instruction=instruction, input=input_context)
    else:
        prompt_start = PROMPT_NO_INPUT_TEMPLATE.format(instruction=in
        struction)
    full_text = prompt_start + output + tokenizer.eos_token
    tokenized_full = tokenizer(full_text, truncation=True, padding=False,
    max_length=MAX_LENGTH)
    tokenized_prompt = tokenizer(prompt_start, truncation=True,
    padding=False, max_length=MAX_LENGTH)
    prompt_length = len(tokenized_prompt["input_ids"])
    labels = copy.deepcopy(tokenized_full["input_ids"])
    for i in range(prompt_length):
        if i < len(labels): labels[i] = -100
    tokenized_full["labels"] = labels
    return tokenized_full

print(f"\n--- Processing Dataset: {DATASET_NAME} ---")
try:
    raw_dataset = load_dataset(DATASET_NAME, split=f"train[:{NUM_SAMPLES_
    PER_DATASET}]")
    tokenized_dataset = raw_dataset.map(
        format_and_tokenize, remove_columns=raw_dataset.column_names
    )
    split_ds = tokenized_dataset.train_test_split(test_size=0.1, seed=42)
    final_datasets = {"train": split_ds["train"], "validation": split_
    ds["test"]}
    print("Dataset processed.")
```

```python
except Exception as e:
    print(f"Error processing dataset: {e}")
    exit()

# --- 3. Load Model with UnslothAI ---
# Unsloth handles quantization and optimizations internally during loading.
print(f"\nLoading model ({MODEL_CHECKPOINT}) with UnslothAI...")
# Determine max_seq_length based on data preprocessing
max_seq_length = MAX_LENGTH
# Determine dtype ("auto", None, torch.float16, torch.bfloat16)
dtype = None # Autodetect
load_in_4bit = True # Enable QLoRA optimizations

try:
    model, tokenizer = FastLanguageModel.from_pretrained(
        model_name = MODEL_CHECKPOINT,
        max_seq_length = max_seq_length,
        dtype = dtype,
        load_in_4bit = load_in_4bit,
        # token = "hf_...", # Add token if using gated models like Llama
    )
    print("Unsloth FastLanguageModel loaded successfully.")
     # Ensure pad token ID is set correctly after potential Unsloth
        modifications
    if tokenizer.pad_token is None:
        tokenizer.pad_token = tokenizer.eos_token
        print(f"Set PAD token to EOS token: {tokenizer.pad_token}")
    # Unsloth might handle model's pad_token_id internally, but check
        if needed
    # if model.config.pad_token_id is None: model.config.pad_token_id =
        tokenizer.pad_token_id

except Exception as e:
    print(f"Error loading model with UnslothAI: {e}")
    print("Ensure UnslothAI is installed correctly for your environment.")
    exit()
```

```python
# --- 4. Apply PEFT (LoRA) Configuration ---
# Unsloth integrates with PEFT. Model is already prepared.
print("\nApplying PEFT LoRA adapters using Unsloth's optimized method...")
try:
    # Use FastLanguageModel.get_peft_model instead of standard get_
      peft_model
    model = FastLanguageModel.get_peft_model(
        model, # Pass the Unsloth model object
        r = LORA_R,
        target_modules = LORA_TARGET_MODULES,
        lora_alpha = LORA_ALPHA,
        lora_dropout = LORA_DROPOUT,
        bias = "none", # Or "all" or "lora_only"
        use_gradient_checkpointing = True, # Recommended by Unsloth
        random_state = 3407, # From Unsloth examples
        max_seq_length = max_seq_length, # Optional but good practice
        use_rslora = False,  # Rank Stable LoRA (experimental)
        loftq_config = None, # LoftQ configuration (experimental)
    )
    print("Unsloth PEFT model created successfully.")
    # Unsloth model object might not have print_trainable_parameters, but
      PEFT is applied.
except Exception as e:
    print(f"Error applying PEFT with Unsloth: {e}")
    exit()

# --- 5. Training Arguments ---
# Arguments are largely the same as standard Trainer
print("\nDefining Training Arguments...")
bf16_supported = is_bfloat16_supported() # Use unsloth's check

training_args = TrainingArguments(
    output_dir=OUTPUT_DIR,
    num_train_epochs=NUM_EPOCHS,
    per_device_train_batch_size=BATCH_SIZE,
    gradient_accumulation_steps=GRADIENT_ACCUMULATION_STEPS,
```

```python
    learning_rate=LEARNING_RATE,
    # Unsloth works with standard optimizers, paged optimizers often
      still good
    optim="paged_adamw_8bit",
    weight_decay=0.01,
    evaluation_strategy="epoch",
    save_strategy="epoch",
    logging_strategy="steps",
    logging_steps=10,
    load_best_model_at_end=True,
    metric_for_best_model="loss",
    fp16=not bf16_supported, # Use fp16 if bf16 not available
    bf16=bf16_supported, # Use bf16 if available
    # gradient_checkpointing is handled by Unsloth's get_peft_model
    push_to_hub=False,
    report_to="none"
)

# --- 6. Data Collator ---
data_collator = DataCollatorForLanguageModeling(tokenizer=tokenizer,
mlm=False)

# --- 7. Initialize Trainer ---
# Use standard Hugging Face Trainer
trainer = Trainer(
    model=model, # Pass the Unsloth PEFT model
    args=training_args,
    train_dataset=final_datasets["train"],
    eval_dataset=final_datasets["validation"],
    tokenizer=tokenizer,
    data_collator=data_collator,
)

# --- 8. Train ---
print("\nStarting Unsloth accelerated fine-tuning...")
# Expect faster training and potentially lower memory usage compared to
standard QLoRA
```

```python
try:
    train_result = trainer.train()
    # Unsloth saves adapters in a compatible format
    # Use trainer.save_model() which internally calls model.save_
      pretrained()
    trainer.save_model(OUTPUT_DIR)
    print(f"Unsloth training finished. Adapter saved to {OUTPUT_DIR}")
    metrics = train_result.metrics
    trainer.log_metrics("train", metrics)
    trainer.save_metrics("train", metrics)
    trainer.save_state()
except Exception as e:
    print(f"Error during Unsloth training: {e}")
    exit()

# --- 9. Evaluate ---
print("\nEvaluating final Unsloth PEFT model...")
try:
    eval_results = trainer.evaluate()
    print("Evaluation Results (Loss):")
    print(eval_results)
    trainer.log_metrics("eval", eval_results)
    trainer.save_metrics("eval", eval_results)
except Exception as e:
    print(f"Error during evaluation: {e}")

# --- End of Recipe ---
```

A strong recommendation to refer to the official UnslothAI GitHub repository
(https://unsloth.ai/) or the documentation for the most up-to-date list of currently
supported and experimental models.

A brief summary stating that it is primarily optimized for transformer-based Causal
Language Models (CLMs) compatible with the Hugging Face ecosystem and that its
benefits are maximized on models with significant training time and memory overhead.

To clearly illustrate the revolutionary impact of Parameter-Efficient Fine-Tuning,
especially when moving from the resource-intensive Full Fine-Tuning (FT) to the highly

optimized QLoRA method, Table 8-1 provides a quantitative comparison of the key metrics. This comparison highlights the dramatic reduction in trainable parameters, VRAM footprint, and the resulting improvements **in training speed, while also considering the mitigation of catastrophic forgetting.**

Table 8-1. *QLoRA: Pushing Efficiency Further with Quantization*

Metric	Full Fine-Tuning (FT)	Standard LoRA (Low-Rank Adaptation)	QLoRA (Quantized LoRA)
Trainable Parameters	100% (All base model weights)	< 1% (Only small A and B matrices)	< 1% (Only small A and B matrices)
Model VRAM Footprint	Highest (Requires VRAM for weights, gradients, optimizer states for all parameters)	High Reduction (Saves space by freezing base model weights, but base model is still but base model is still loaded in >= 16-bit precision)	Maximal Reduction (Base model is loaded in 4-bit precision, drastically reducing memory load)
VRAM Requirement (Gemma-2B est.)	Often > 30 GB	Often > 24 GB (as per Recipe 1 note)	Typically 12–16GB (or less, making it consumer-grade accessible)
Training Speed	Baseline/Slowest	Faster (Fewer parameters to calculate gradients for)	Fast (Faster than FT; further improved by memory efficiency and smaller batch times)
Risk of Catastrophic Forgetting	Present	Low (Base weights are frozen)	Low (Base weights are frozen)

While the techniques discussed are all effective, choosing the right PEFT method depends on your specific hardware constraints and desired outcome. The following decision framework (Table 8-2) summarizes the primary recommendations based on the core trade-offs: speed, memory consumption (VRAM), and implementation complexity.

Table 8-2. *Comparison of PEFT Strategies: LoRA, QLoRA, and UnslothAI*

Condition	Recommended Method	Rationale (Based on Document Content)
Maximal Speed & Minimal VRAM	UnslothAI (Accelerated QLoRA)	UnslothAI uses highly optimized kernels to significantly speed up QLoRA/LoRA training and further reduce memory usage. It is the best choice when pushing for the fastest possible results and largest batch sizes on consumer-grade GPUs. (Requires model to be on UnslothAI's supported list)
Significant VRAM Reduction & Consumer Hardware	Standard QLoRA (4-bit)	This is the primary recommendation for accessing large models (like Gemma-2B) on consumer hardware (typically 12–16GB VRAM). It is a standard, highly effective method within the Hugging Face ecosystem and provides a maximal reduction in the base model's memory footprint via 4-bit quantization.
Moderate VRAM Reduction & Pure HF Implementation	Standard LoRA	Use this if you have a larger VRAM budget (Gemma-2B est. > 24GB) but still want the parameter-efficiency and catastrophic forgetting mitigation benefits of PEFT. It is a simpler implementation without the complexity of 4-bit quantization, but it is much more resource intensive than QLoRA.
Model Incompatibility	Standard LoRA/ QLoRA	If your specific model architecture is not yet supported by UnslothAI's specialized kernels, standard LoRA or QLoRA (using bitsandbytes and peft) remains the reliable, fully compatible implementation within the general Hugging Face ecosystem.

Chapter Summary

In your journey to adapt large language models, you saw the power of full fine-tuning, which meticulously updated every parameter for a new task. However, as you may have experienced, this approach requires significant computational resources, vast GPU memory, and time, often making the adaptation of massive models out of reach. This chapter has explored Parameter-Efficient Fine-Tuning (PEFT), a transformative set of techniques for adapting large language models, showing you how to bypass the high computational cost of full fine-tuning. We hope you've learned that PEFT methods work by drastically reducing resource requirements: we achieve this by freezing the base model and training only a small fraction of parameters. You saw that the dominant technique, LoRA, achieves this efficiency through low-rank adaptation. Furthermore, we showed you how QLoRA enhances this with 4-bit quantization, enabling you to access and adapt large models even on consumer hardware. This powerful efficiency was quantitatively demonstrated for you by comparing key metrics against Full Fine-Tuning in Table 8-1. Finally, we provided you with a clear decision framework in Table 8-2 that summarizes the primary recommendations based on factors such as speed, VRAM, and implementation complexity, and we observed that specialized libraries like UnslothAI can deliver significant speed and memory improvements over standard implementations. Building on this foundation, the next chapter will focus on augmenting model training with synthetic data to further enhance your fine-tuning capabilities.

Augmenting with Synthetic Data

Having explored both comprehensive and parameter-efficient methods for fine-tuning, we now address a common bottleneck that underlies all adaptation techniques: the availability of data. Sometimes, the high-quality, domain-specific data needed for a perfect fine-tuning process is simply unavailable. You might lack sufficient examples for a niche task, need to cover rare instruction types, or face privacy constraints that prevent the use of real user data. In such situations, we rely on a powerful and increasingly crucial technique: **Synthetic Data Generation**. This chapter will explore how to create data artificially, using other AI models as a tool, allowing us to augment, bootstrap, or even replace traditional datasets to achieve our fine-tuning goals.

Introduction to Synthetic Data

In the context of LLMs, synthetic data refers to text—be it instructions, responses, or entire conversations—generated programmatically, most often by another, more powerful AI model (a "teacher" model). This approach has become a cornerstone of modern AI development, addressing several critical challenges:

- **Data Scarcity:** It overcomes the lack of real-world data for specialized domains or languages.

- **Data Augmentation:** It increases the volume and diversity of an existing dataset, improving model robustness.

- **Privacy Preservation:** It enables the creation of realistic yet artificial data that mimics sensitive information without disclosing it.

© Bharath Kumar Bolla, Kalpa Subbaiah and Sashi Kiran Kaata 2026
B. K. Bolla et al., *Large Language Model Recipes*, https://doi.org/10.1007/979-8-8688-2607-8_9

- **Cost Reduction:** Generating data via an API can be significantly cheaper and faster than large-scale human annotation.

The impact of high-quality synthetic data has been demonstrated in several recent state-of-the-art models. Microsoft's **Phi** series of small language models, for instance, achieved surprising performance by being trained on a heavily curated dataset composed almost entirely of synthetic, "textbook-quality" data generated by larger models. Similarly, Google has indicated that the development of its powerful **Gemini** models involved using high-quality data from its own best models to help train subsequent versions, creating a cycle of improvement.

The most common and effective method for creating this data is to leverage a highly capable existing LLM as a "teacher."

Generating Synthetic Data with a Teacher LLM

The core idea is to use a state-of-the-art model (like GPT-4, Claude 3, or Gemini) to generate new training examples. One of the most straightforward applications is generating question-answer pairs from a given document, instantly creating a dataset suitable for Q&A fine-tuning. The Figure 9-1 illustrates the process of generating question-answer pairs using a multi-model pipeline where a generator LLM creates candidates and a judge LLM ranks and selects the best ones for the final synthetic dataset.

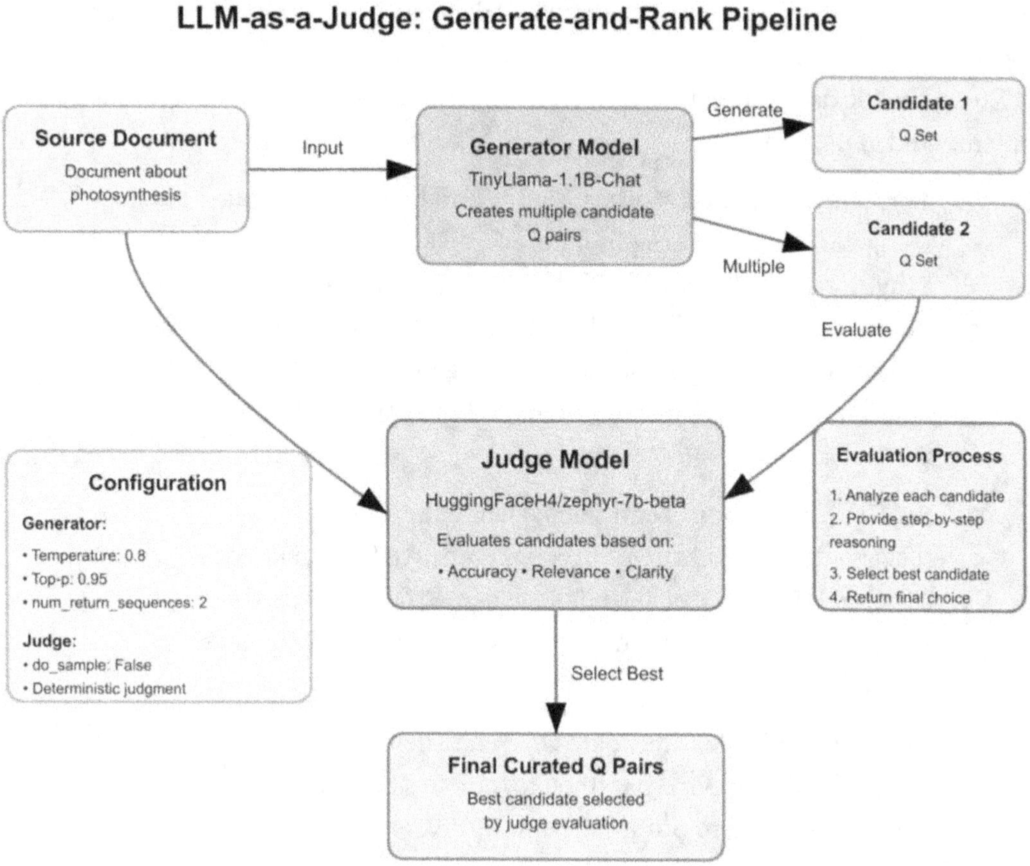

LLM-as-a-Judge: Generate-and-Rank Pipeline

Figure 9-1. LLM-As-a-Judge

Recipe 1: Generating Question-Answer Pairs from a Document

Goal: Use a multi-model, generate-and-rank pipeline to create high-quality, document-grounded synthetic Q&A data.

This recipe demonstrates a **"Generate and Rank"** pipeline for generating high-quality instruction-following data (specifically, question-answer pairs) directly from a source document. It implements a fully automated quality assurance step by leveraging a two-model system: a faster **Generator** (the "Student") for generating diverse candidate outputs and a stronger **Judge** (the "Teacher") for evaluating and curating the output against strict criteria. This separation ensures the final dataset is of higher quality and mitigates the risk of low-quality or hallucinated data contaminating the fine-tuning process.

The synthetic data generation workflow for this recipe follows these general steps, as implemented in the code:

1. **Configure and Load Models:** Define and load two separate LLMs using the `transformers` pipeline: a fast Generator Model (e.g., `TinyLlama/TinyLlama-1.1B-Chat-v1.0`) for bulk creation and a powerful Judge Model (e.g., `HuggingFaceH4/zephyr-7b-beta`) for reasoning and critique. Resource management parameters such as torch.bfloat16 and device_map="auto" are used to optimize GPU utilization.

2. **Set Up Prompts and Templates:** Define the source text (document), the `generator_prompt_template` (which forces a structured JSON output), and the `judge_prompt_template` (which sets the explicit evaluation criteria: **Accuracy, Relevance, and Clarity**).

3. **Generate Multiple Candidate QA Sets:** The Generator model is called with sampling parameters (`num_return_sequences=2, do_sample=True`) to produce multiple, diverse sets of Q&A pairs. Regular expressions are used to reliably extract the clean JSON list from the model's textual response.

4. **Evaluate and Select Best Set:** The Judge Model is prompted with the source document and all candidate sets. By setting do_sample=False, the Judge's selection process is deterministic based on its best reasoning and adherence to the defined quality criteria.

5. **Extract Curated Dataset:** A final regular expression extracts the **"Final Choice"** (the single, best JSON list) from the Judge's output, which is then parsed to form the clean, curated synthetic dataset.

What to Expect

This process yields a small, highly curated dataset of question-answer pairs that are factually grounded in the source text and pre-filtered for quality. The key benefit of this approach is that it automates the quality control of synthetic data using a more capable "teacher" model, resulting in a higher-quality, task-specific dataset that is essential for successful fine-tuning.

```python
# --- Recipe 1: Generating Question-Answer Pairs with a Judge Model ---
# Goal: Use a multi-model, generate-and-rank pipeline to create high-
quality synthetic data.
# Prerequisites: pip install transformers torch accelerate bitsandbytes
# Note: This procedure loads TWO models. While smaller models are used here
to reduce VRAM,
#        it can still be resource-intensive (>16GB recommended).

import torch
from transformers import pipeline, AutoTokenizer
import json
import re

# 1. --- Configuration ---
# A small, fast model to generate multiple candidate answers
GENERATOR_MODEL_ID = "TinyLlama/TinyLlama-1.1B-Chat-v1.0"
# A separate, powerful model (small but strong reasoning) to evaluate the
candidates
JUDGE_MODEL_ID = "HuggingFaceH4/zephyr-7b-beta"

# 2. --- Setup the Generator and Judge Model Pipelines ---
print(f"Loading GENERATOR model: {GENERATOR_MODEL_ID}")
try:
    generator_tokenizer = AutoTokenizer.from_pretrained(GENERATOR_MODEL_ID)
    generator_pipe = pipeline(
        "text-generation",
        model=GENERATOR_MODEL_ID,
        tokenizer=generator_tokenizer,
        torch_dtype=torch.bfloat16,
        device_map="auto",
    )
    print("Generator model pipeline loaded successfully.")
except Exception as e:
    print(f"Error loading generator model pipeline: {e}")
```

```python
    exit()

print(f"\nLoading JUDGE model: {JUDGE_MODEL_ID}")
try:
    judge_tokenizer = AutoTokenizer.from_pretrained(JUDGE_MODEL_ID, trust_
    remote_code=True)
    judge_pipe = pipeline(
        "text-generation",
        model=JUDGE_MODEL_ID,
        tokenizer=judge_tokenizer,
        torch_dtype=torch.bfloat16,
        device_map="auto",
        trust_remote_code=True,
    )
    print("Judge model pipeline loaded successfully.")
except Exception as e:
    print(f"Error loading judge model pipeline: {e}")
    exit()

# 3. --- Define the Document and Prompt Templates ---
document = """
Photosynthesis is a process used by plants, algae, and certain bacteria
to convert light energy into chemical energy, through a process that
converts carbon dioxide and water into sugars (glucose) and oxygen. This
process is crucial for life on Earth as it produces most of the oxygen in
the atmosphere. The chemical equation for photosynthesis is 6CO2 + 6H2O →
C6H12O6 + 6O2. Chlorophyll is the primary pigment used in photosynthesis;
it absorbs blue and red light and reflects green light, which is why plants
appear green.
"""

generator_prompt_template = """
Given the following document, generate three distinct question-answer
pairs based on its content. The questions should be insightful and the
answers should be accurate and concise. Format the output as a list of JSON
objects, where each object has a "question" and an "answer" key.
```

```
Document:
\"\"\"
{document}
\"\"\"

JSON Output:
"""

judge_prompt_template = """
You are an expert evaluator. Your task is to analyze multiple sets of
synthetically generated question-answer pairs based on a source document.
Evaluate them based on the following criteria:
1.  **Accuracy**: Is the answer factually correct according to the
document?
2.  **Relevance**: Does the question directly relate to a key concept in
the document?
3.  **Clarity**: Are the question and answer easy to understand?

Below are the source document and the candidate QA sets.

Source Document:
\"\"\"
{document}
\"\"\"

---

Candidate QA Sets:
{candidate_sets}
---

First, provide a brief step-by-step reasoning for your choice, explaining
which set is superior and why.
Finally, on a new line, state your final choice by reprinting the single
best JSON list.

Reasoning:

Final Choice:
"""
```

```python
# 4. --- Generate Multiple Candidate QA Sets ---
prompt = generator_prompt_template.format(document=document)
print("\nGenerating multiple candidate synthetic data sets...")
candidate_responses = []
try:
    # Generate two different sets of responses
    outputs = generator_pipe(
        prompt,
        max_new_tokens=300,
        num_return_sequences=2, # Generate two candidates
        do_sample=True,
        temperature=0.8,
        top_p=0.95,
        eos_token_id=generator_tokenizer.eos_token_id
    )

    for i, out in enumerate(outputs):
        generated_text = out['generated_text']
        response_part = generated_text[len(prompt):]
        # Find the JSON list within the response
        json_match = re.search(r'\[.*\]', response_part, re.DOTALL)
        if json_match:
            candidate_responses.append(json_match.group(0))
            print(f"\n--- Candidate {i+1} ---")
            print(candidate_responses[-1])

    if not candidate_responses:
        raise ValueError("No valid JSON lists were generated by the
        generator model.")

except Exception as e:
    print(f"\nAn error occurred during candidate generation: {e}")
    exit()

# 5. --- Use the Judge Model to Rank and Select the Best Set ---
print("\n--- Submitting candidates to the Judge Model for evaluation ---")
```

```python
# Format the candidates for the judge prompt
formatted_candidates = ""
for i, candidate_json in enumerate(candidate_responses):
    formatted_candidates += f"Candidate {i+1}:\n{candidate_json}\n\n"

judge_prompt = judge_prompt_template.format(
    document=document,
    candidate_sets=formatted_candidates
)

try:
    judge_outputs = judge_pipe(
        judge_prompt,
        max_new_tokens=512,
        do_sample=False, # We want a deterministic judgment
        eos_token_id=judge_tokenizer.eos_token_id
    )

    judge_response = judge_outputs[0]['generated_text'][len(judge_prompt):]

    print("\n--- Judge's Full Response ---")
    print(judge_response)

    # --- Extract the Final Curated Dataset ---
    final_choice_match = re.search(r'Final Choice:\s*(\[.*\])', judge_
response, re.DOTALL)

    if final_choice_match:
        final_json_str = final_choice_match.group(1)
        parsed_json = json.loads(final_json_str)
        pretty_json = json.dumps(parsed_json, indent=2)

        print("\n--- Final Curated QA Pairs (Selected by Judge) ---")
        print(pretty_json)
    else:
        print("\n--- Could not parse the final choice from the judge's
output ---")

except Exception as e:
    print(f"\nAn error occurred during judgment: {e}")
```

Success Criteria

The success of a single QA pair is explicitly defined by the **Judge Model's three evaluation criteria**:

- **Accuracy:** Is the answer factually correct according to the source document? (This prevents hallucination.)

- **Relevance:** Does the question directly relate to a key concept in the document? (This ensures task specificity.)

- **Clarity:** Are the question and answer easy to understand? (This ensures data quality for the fine-tuning process.)

"Good" Output Look Like: A JSON list of Q&A pairs that have been validated by the powerful Judge Model as meeting all three criteria.

Building on this, we can use a teacher model not just to answer questions but to invent entirely new instructions. This technique, known as **Self-Instruct**, is a powerful method for bootstrapping a diverse instruction-following dataset from just a few seed examples.

The Self-Instruct Method

The Self-Instruct loop is a scalable process for creating a large and diverse instruction dataset:

1. **Start with Seed Examples:** Begin with a small set of human-written, diverse instructions.

2. **Generate New Instructions:** Prompt the teacher LLM with these seed examples to generate new, similar yet distinct instructions.

3. **Generate Responses:** Prompt the teacher LLM again to generate high-quality responses for each newly created instruction.

4. **Filter for Quality:** Rigorously filter the generated pairs to remove low-quality, repetitive, or incorrect examples.

5. **Repeat:** Add the newly high-quality pairs back to the seed set and then repeat the process.

This method was famously used to create the dataset for the original Stanford Alpaca model and has become a standard technique for dataset creation. The Figure 9-2 depicts the Self-Instruct loop, a method where a teacher LLM generates new, diverse instructions based on a small set of seed examples, allowing for scalable dataset creation.

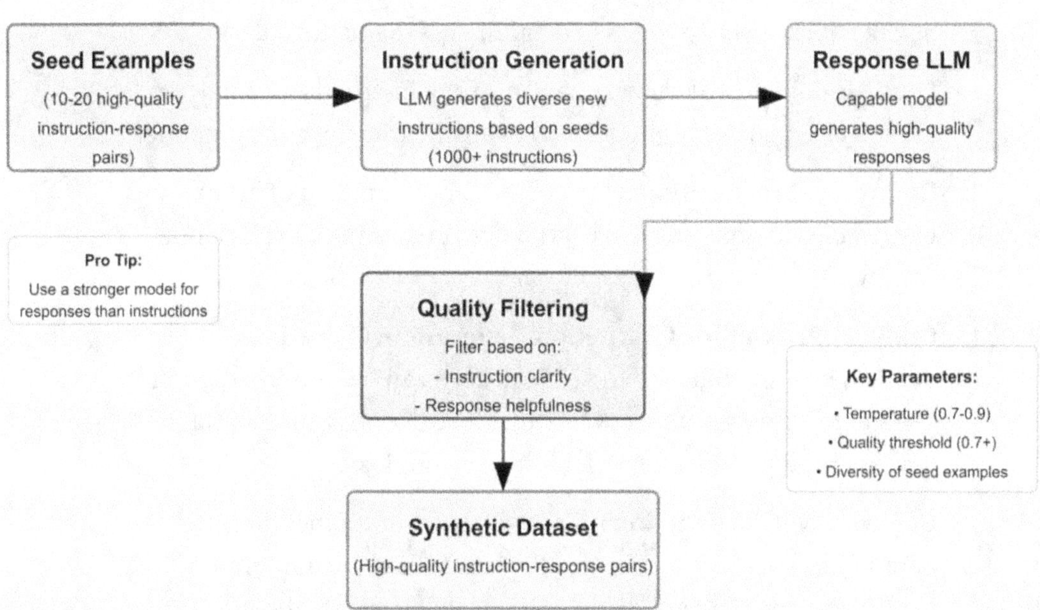

Figure 9-2. *Self-Instruct Self-Generated Data Approach*

Recipe 2: Generating New Instructions (Self-Instruct Style)

Goal: Use a teacher LLM to generate new instructions based on a few seed examples.

This procedure demonstrates the **Self-Instruct** method, a scalable and powerful technique for bootstrapping a large, diverse instruction-following dataset from a small initial set of examples. The core idea is to leverage a highly capable, instruction-tuned LLM (the "Teacher") to invent new task instructions that are similar in nature but distinct from the original examples. This method is critical for quickly expanding the diversity of a fine-tuning dataset without requiring extensive human annotation.

The self-instruct workflow for this recipe follows these general steps, as implemented in the code:

1. **The Self-Instruct Loop:** The complete method involves starting with a small **Seed Examples** set, using the teacher LLM to **Generate New Instructions**, prompting the teacher LLM again to **Generate Responses** for those new instructions, and then **Filtering for Quality** before repeating the process to continually expand the dataset.

2. **Load Model and Tokenizer:** Load a strong instruction-tuned model (e.g., google/gemma-2b-it) using the transformers pipeline.

3. **Define Seed Instructions:** Create a small, diverse set of human-written instructions covering various task types (e.g., explanation, writing, and coding).

4. **Craft the Prompt for Instruction Generation:** Construct a detailed prompt that asks the LLM to act as an "expert instruction generator" and produce new, distinct instructions that maintain the topic and task diversity of the seed examples.

5. **Generate New Instructions:** Call the generation pipeline with parameters such as do_sample=True and a higher temperature (e.g., 0.8) to maximize the originality and diversity of the new instructions.

6. **Post-processing:** Extract the generated, numbered list of new instructions from the model's text output for subsequent filtering and response generation.

What to Expect

This process yields a list of novel, clear, and actionable instructions that significantly expand the task coverage of your dataset. It is important to note that this recipe focuses solely on generating *instructions*. The next step in the full Self-Instruct loop, as outlined in the document, is to make another call to a powerful LLM to generate high-quality responses for each newly created instruction.

```
# --- Recipe 2: Generating Instructions (Self-Instruct Style) ---
# Goal: Use an LLM to generate new, diverse instructions based on seed
examples.
```

```python
# Method: Prompts an instruction-tuned LLM with examples to generate more
instructions.
# Note: This only generates the *instructions*, not the corresponding
outputs.
#        Quality depends heavily on the LLM and the seed examples.

import torch
from transformers import pipeline

# --- Configuration ---
# Use a strong instruction-tuned model if possible
MODEL_ID = "google/gemma-2b-it" # Or Mistral-Instruct, etc.
MAX_NEW_TOKENS_INSTR = 300 # Max tokens for the list of new instructions

# Use GPU if available
device_index = 0 if torch.cuda.is_available() else -1
dtype = torch.bfloat16 if torch.cuda.is_available() and torch.cuda.is_bf16_
supported() else torch.float32

# --- 1. Load Model and Tokenizer (via Pipeline) ---
print(f"Loading pipeline for model: {MODEL_ID}")
try:
    generator = pipeline(
        "text-generation",
        model=MODEL_ID,
        tokenizer=MODEL_ID,
        torch_dtype=dtype,
        device=device_index,
    )
    if generator.tokenizer.pad_token is None: generator.tokenizer.pad_token
    = generator.tokenizer.eos_token
    if generator.model.config.pad_token_id is None: generator.model.config.
    pad_token_id = generator.tokenizer.pad_token_id
    print("Pipeline loaded.")
except Exception as e:
    print(f"Error loading pipeline: {e}")
    exit()
```

```python
# --- 2. Define Seed Instructions ---
# Provide a diverse set of examples to guide the generation
seed_instructions = [
    "Explain the difference between supervised and unsupervised learning.",
    "Write a short email inviting a colleague to a project meeting.",
    "Generate a list of 5 creative names for a new coffee shop.",
    "What are the main benefits of using renewable energy sources?",
    "Provide Python code to read a CSV file into a pandas DataFrame.",
]

# --- 3. Craft the Prompt for Instruction Generation ---
# The prompt asks the model to act as an instruction generator
# and create new instructions similar to, but distinct from, the seeds.
prompt = f"""You are an expert instruction generator. Your task is to
create a list of 5 new, diverse instructions that are different from the
examples provided below, but cover a similar range of topics and task types
(e.g., explanation, writing, brainstorming, coding, Q&A).

Do not repeat the examples. Ensure the new instructions are clear and
actionable.

Examples of existing instructions:
\"\"\"
- {seed_instructions[0]}
- {seed_instructions[1]}
- {seed_instructions[2]}
- {seed_instructions[3]}
- {seed_instructions[4]}
\"\"\"

Generate 5 new instructions below, formatted as a numbered list:
1. """ # Start the list for the model

print("--- Generating New Instructions ---")
print(f"Prompt being sent to model:\n{prompt}")

# --- 4. Generate New Instructions ---
try:
```

```python
    outputs = generator(
        prompt,
        max_new_tokens=MAX_NEW_TOKENS_INSTR,
        num_return_sequences=1,
        do_sample=True,
        temperature=0.8, # Higher temperature for more diversity
        top_p=0.95,
        pad_token_id=generator.tokenizer.eos_token_id
    )

    generated_text = outputs[0]['generated_text']
    # Extract the generated list part
    instruction_list_str = "1. " + generated_text.split("Generate 5 new
    instructions below, formatted as a numbered list:\n1. ")[-1].strip()

    print("\n--- Generated Instructions ---")
    print(instruction_list_str)

    # --- 5. Post-processing (Optional) ---
    # Split the string into a list of instructions
    generated_instructions = [line.split('. ', 1)[-1] for line in
    instruction_list_str.split('\n') if line.strip() and line[0].isdigit()]
    print("\n--- Parsed Instructions ---")
    print(generated_instructions)
    # Next step would be to filter these and then generate outputs for them
    (using another LLM call).

except Exception as e:
    print(f"Error during instruction generation: {e}")

# --- End of Recipe ---
```

Data Augmentation Through Transformation

Beyond generating entirely new data points, we can also **augment** an existing
dataset by applying transformations to its examples. This is a powerful and cost-
effective way to increase the size and diversity of a small, high-quality seed dataset.
By creating variations of existing data—such as paraphrasing, back-translation, or

changing the length and style—we can expose the model to a wider range of linguistic patterns, improving its robustness and generalization. The Figure 9-3 shows the Data Augmentation & Transformation Approach, detailing how a small original dataset is expanded using techniques like paraphrasing, back-translation, and style transfer, followed by quality verification.

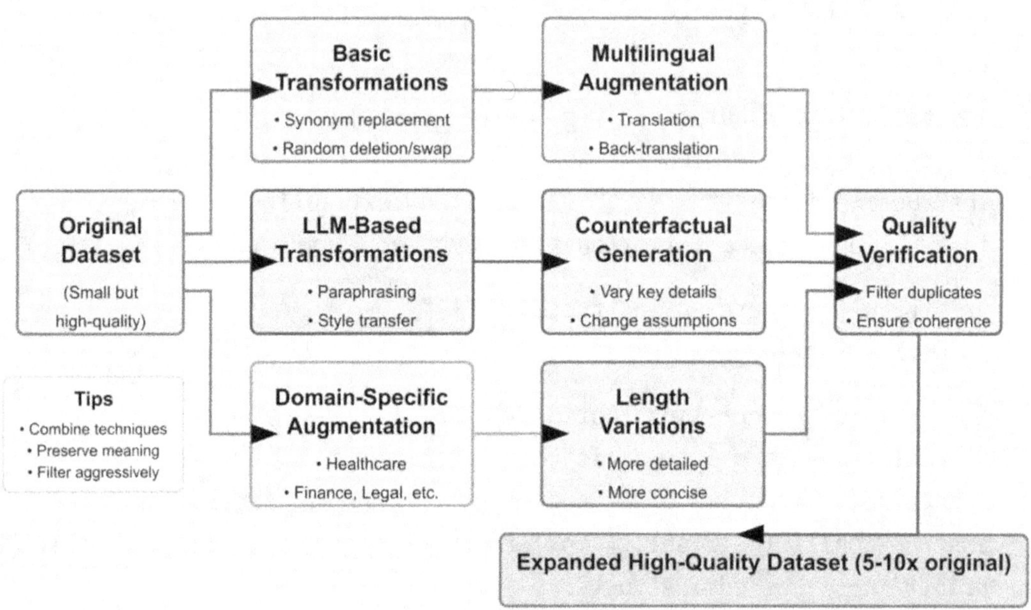

Figure 9-3. *Data Augmentation and Transformation Approach. This diagram shows how a small, high-quality original dataset can be expanded through various techniques. These include basic transformations (like synonym replacement), advanced LLM-based transformations (like paraphrasing and style transfer), and other methods like back-translation and counterfactual generation, all followed by a quality verification step*

Note on Scaling and Production Readiness

For production-scale Self-Instruct pipelines (generating 1,000+ instructions), the process must be automated beyond this example. Key professional practices include using a faster, cheaper LLM for the bulk instruction/response generation; implementing parallel API calls; and, crucially, adding automated quality filters like embedding-based deduplication, perplexity checks, and toxicity screening to ensure the final dataset's integrity.

Recipe 3: Data Augmentation via Transformation

Goal: Augment a small dataset using LLM-based paraphrasing, back-translation, and length variation.

This procedure demonstrates the **Data Augmentation Through Transformation** method, a powerful and cost-effective way to increase the size and diversity of a small, high-quality seed dataset. The core idea is to create new training examples by systematically applying various linguistic transformations—like paraphrasing, back-translation, and style transfer—to existing data points. This exposes the fine-tuning model to a wider range of linguistic patterns, significantly improving its robustness and generalization.

The data augmentation workflow for this recipe follows these general steps, as implemented in the code:

1. **Setup Augmentation Models:** Load multiple models, including translation pipelines (e.g., Helsinki-NLP/opus-mt-en-de) for multilingual augmentation (back-translation) and a text generation LLM (e.g., TinyLlama/TinyLlama-1.1B-Chat-v1.0) for more sophisticated, LLM-based transformations.

2. **Define Original Data Point:** Select an initial, high-quality text snippet (e.g., a sentence from a financial report) from the existing dataset to be the basis for all new examples.

3. **Apply LLM-Based Transformations:** Use the augmenter LLM to create sophisticated variations, including **Paraphrasing, Length Variations** (summarize/elaborate), **Counterfactual Generation** (opposite meaning), and **Domain-Specific Style Transfer** (e.g., to a legal style).

4. **Apply Multilingual and Basic Transformations:** Perform **Back-Translation** (e.g., English ➤ German ➤ English) to introduce linguistic noise, and apply simple **Rule-Based Transformations** like Synonym Replacement.

5. **Display Augmented Samples:** Present the final dictionary of the original text and all generated variations, ready for inclusion in the final training dataset after an optional quality verification step.

What to Expect

This process yields a high-volume, diverse set of linguistic variations for a single data point, significantly expanding the dataset's coverage of phrasing, style, and length without requiring new human-annotated content. This is a critical step for improving a model's ability to generalize across different phrasings and user inputs.

```python
# --- Procedure 3: Data Augmentation via Transformation ---
# Goal: Augment a small dataset using a variety of LLM-based and rule-based
transformation techniques.
# Prerequisites: pip install transformers[sentencepiece] torch accelerate
# Note: This procedure uses multiple models and can be resource-intensive.

import torch
from transformers import pipeline
import random

# 1. --- Setup Augmentation Models ---
print("Setting up augmentation models...")
# Use translation pipelines for multilingual augmentation (back-
translation)
try:
    translator_en_de = pipeline("translation_en_to_de", model="Helsinki-
    NLP/opus-mt-en-de", device_map="auto")
    translator_de_en = pipeline("translation_de_to_en", model="Helsinki-
    NLP/opus-mt-de-en", device_map="auto")
    print("Translation models loaded.")
except Exception as e:
    print(f"Could not load translation models: {e}")
    translator_en_de, translator_de_en = None, None

# Use a text generation model for more sophisticated transformations
try:
    augmenter_pipe = pipeline(
        "text-generation",
        model="TinyLlama/TinyLlama-1.1B-Chat-v1.0",
        torch_dtype=torch.bfloat16,
        device_map="auto"
    )
```

```
    print("Augmenter LLM loaded.")
except Exception as e:
    print(f"Could not load augmenter LLM: {e}")
    augmenter_pipe = None

# 2. --- Original Data Point ---
original_text = "The financial report indicates a significant increase in
quarterly revenue due to strong sales in the technology sector."
print(f"\nOriginal Text:\n\"{original_text}\"")

# Dictionary to store all generated variations
augmented_samples = {"Original": original_text}

# 3. --- Apply Transformations ---

# a) LLM-Based Paraphrasing
print("\n--- Applying LLM-Based Paraphrasing ---")
if augmenter_pipe:
    paraphrase_prompt = f"&lt;|system|&gt;\nYou are a helpful assistant.</s>\
    n&lt;|user|&gt;\nParaphrase the following sentence while preserving its core
    meaning: \"{original_text}\"</s>\n&lt;|assistant|&gt;\n"
    paraphrased_text = augmenter_pipe(paraphrase_prompt, max_new_
    tokens=70, do_sample=True, temperature=0.7)[0]['generated_text'].
    split("&lt;|assistant|&gt;")[-1].strip()
    augmented_samples["Paraphrased"] = paraphrased_text

# b) Multilingual Augmentation (Back-Translation)
print("\n--- Applying Back-Translation (English -> German -> English) ---")
if translator_en_de and translator_de_en:
    translated_to_de = translator_en_de(original_text)[0]
    ['translation_text']
    back_translated_text = translator_de_en(translated_to_de)[0]
    ['translation_text']
    augmented_samples["Back-Translated"] = back_translated_text

# c) Length Variations (Summarize and Elaborate)
print("\n--- Applying Length Variations ---")
if augmenter_pipe:
```

```python
    summarize_prompt = f"&lt;|system|&gt;\nYou are a helpful assistant.</s>\
n&lt;|user|&gt;\nSummarize this sentence in fewer words: \"{original_text}\"</
s>\n&lt;|assistant|&gt;\n"
    summarized_text = augmenter_pipe(summarize_prompt, max_new_tokens=25,
    do_sample=False)[0]['generated_text'].split("&lt;|assistant|&gt;")[-1].strip()
    augmented_samples["Summarized"] = summarized_text

    elaborate_prompt = f"&lt;|system|&gt;\nYou are a helpful assistant.</s>\
n&lt;|user|&gt;\nElaborate on this sentence, adding more specific but
    plausible details: \"{original_text}\"</s>\n&lt;|assistant|&gt;\n"
    elaborated_text = augmenter_pipe(elaborate_prompt, max_new_
    tokens=80, do_sample=True, temperature=0.8)[0]['generated_text'].
    split("&lt;|assistant|&gt;")[-1].strip()
    augmented_samples["Elaborated"] = elaborated_text

# d) Counterfactual Generation
print("\n--- Applying Counterfactual Generation ---")
if augmenter_pipe:
    counterfactual_prompt = f"&lt;|system|&gt;\nYou are a helpful assistant.</
    s>\n&lt;|user|&gt;\nRewrite the following sentence to describe the opposite
    scenario (a negative outcome): \"{original_text}\"</s>\n&lt;|assistant|&gt;\n"
    counterfactual_text = augmenter_pipe(counterfactual_prompt, max_new_
    tokens=70, do_sample=True, temperature=0.7)[0]['generated_text'].
    split("&lt;|assistant|&gt;")[-1].strip()
    augmented_samples["Counterfactual"] = counterfactual_text

# e) Domain-Specific Augmentation (Style Transfer)
print("\n--- Applying Domain-Specific Style Transfer (to Legal) ---")
if augmenter_pipe:
    style_transfer_prompt = f"&lt;|system|&gt;\nYou are a helpful assistant.</
    s>\n&lt;|user|&gt;\nRewrite the following sentence in a formal, legal style:
    \"{original_text}\"</s>\n&lt;|assistant|&gt;\n"
    legal_style_text = augmenter_pipe(style_transfer_prompt, max_new_
    tokens=80, do_sample=True, temperature=0.5)[0]['generated_text'].
    split("&lt;|assistant|&gt;")[-1].strip()
    augmented_samples["Legal Style"] = legal_style_text
```

```python
# f) Basic Transformations (Synonym Replacement)
print("\n--- Applying Basic Transformation (Synonym Replacement) ---")
def synonym_replacement(sentence, synonyms, n=1):
    words = sentence.split()
    new_words = words.copy()
    replaceable_words = [word for word in words if word.lower().strip(".,")
    in synonyms]
    if not replaceable_words:
        return sentence

    for _ in range(n):
        word_to_replace = random.choice(replaceable_words)
        # Preserve punctuation
        punctuation = ''
        clean_word = word_to_replace
        if not word_to_replace[-1].isalnum():
            punctuation = word_to_replace[-1]
            clean_word = word_to_replace[:-1]

        synonym = random.choice(synonyms[clean_word.lower()])

        for i, word in enumerate(new_words):
            if word == word_to_replace:
                new_words[i] = synonym + punctuation
                break
    return " ".join(new_words)

synonym_dict = {
    "report": ["statement", "summary", "document"],
    "significant": ["notable", "substantial", "major"],
    "increase": ["growth", "rise", "expansion"],
    "strong": ["robust", "powerful", "vigorous"]
}
synonym_replaced_text = synonym_replacement(original_text, synonym_
dict, n=2)
```

```python
augmented_samples["Synonym Replaced"] = synonym_replaced_text

# 4. --- Display All Augmented Samples ---
print("\n\n=============================================")
print("        Data Augmentation Results            ")
print("=============================================")
for augmentation_type, text in augmented_samples.items():
    print(f"\n--- {augmentation_type} ---")
    print(f"\"{text}\"")
print("\n=============================================")
```

Advanced Augmentation: Production Considerations

Moving a synthetic data pipeline to production involves more than just running the recipes; it requires continuous monitoring and strategic decision-making. The following are the critical questions that must be addressed for any large-scale augmentation effort:

- **Necessity vs. Waste:** Augmentation is necessary when performance plateaus due to a lack of diversity or volume in the **real-world edge cases** (e.g., rare instructions and specific terminology). It is wasteful when it simply generates redundant or low-quality paraphrases of existing, well-represented data points, which adds cost without improving generalization.

- **Augmentation Saturation:** To determine "how much is enough," track the model's performance on a held-out, human-curated validation set. The saturation point is reached when adding more augmented data provides diminishing returns or when the cost of generating new data exceeds the benefit of the marginal performance gain.

- **Measuring Success:** Success is not measured by the quantity of data, but by the performance lift on the target task. Key metrics to track are

- **External Validation Score:** The score of the fine-tuned model on a clean, real-world test set

- **Internal Quality Metrics:** The percentage of generated data points that pass internal filters (like the Judge Model's criteria, deduplication, and perplexity checks)

- **Debugging Performance Degradation:** When augmented data hurts performance (often called "synthetic data drift"), the problem is almost always poor data quality. The solution is to immediately increase the strictness of the automated filtering: raise the **deduplication threshold**, lower the acceptable **perplexity score**, or transition to a more powerful **Judge Model** for quality curation.

Chapter Summary

This chapter has comprehensively explored the powerful and critical technique of augmenting your fine-tuning datasets with synthetic data. You learned that this approach is essential for overcoming data scarcity, a common challenge in specialized domains, and that it has been a key enabler of recent state-of-the-art models like the Phi series. We guided you through the primary method of leveraging a powerful teacher LLM to generate new examples, detailing two crucial recipes: first, the "Generate and Rank" pipeline to create document-grounded, high-quality question-answer pairs using a Judge Model for rigorous quality assurance and, second, the "Self-Instruct" loop, which shows you how to bootstrap a large, diverse set of novel instructions from just a few seed examples.

Furthermore, you discovered the efficiency of Data Augmentation through Transformation, a cost-effective way to increase diversity by paraphrasing, back-translation, and style transfer on existing data points. Critically, we established that raw synthetic data must always undergo rigorous filtering, using metrics such as deduplication and perplexity checks, to ensure its quality, as low-quality data can easily degrade your model's performance. By mastering these generation and filtering techniques, you are now equipped to create high-quality, bespoke datasets that will significantly enhance the capabilities and robustness of your fine-tuned models. With your model specialized and your data augmented, the next chapter will prepare you for the final step: production, by covering the essential techniques for advanced quantization techniques to optimize your fine-tuned artifacts for efficient deployment and inference.

PART IV

Optimization, Serving, and Evaluation

Model Quantization

Having fine-tuned our model and augmented its training data, we arrive at the final stage of optimization before deployment. Large Language Models, by their nature, are computationally intensive. They consume a significant amount of memory (VRAM and RAM) and can be slow during inference, making deployment challenging, especially on resource-constrained hardware. This chapter introduces **Quantization**, a powerful set of techniques designed to shrink a model's size and accelerate its performance, making it leaner and more efficient for real-world applications.

The Need for Optimization: An Introduction to Quantization

The core challenge with deploying LLMs lies in their use of high-precision floating-point numbers (like 32-bit FP32 or 16-bit BFloat16/Float16). These formats require substantial memory and computational power. A 7-billion parameter model in FP16 needs at least 14GB of VRAM just to load its weights, plus more for the intermediate calculations (activations) during inference.

Quantization provides the solution. The process involves converting a model's weights, and sometimes its activations, from high-precision formats to lower-precision numerical types, such as 8-bit integers (INT8) or 4-bit integers (INT4). This conversion has two primary benefits:

Numeric Format Comparison for Model Quantization

FP32 S	8-bit Exponent	23-bit Mantissa
FP16 S	5-bit Exp · 10-bit Mant	50% size reduction from FP32
INT8 S	7-bit Int	75% size reduction from FP32
NF4 S E 3b		87.5% size reduction from FP32
FP4 S E 2b		87.5% size reduction from FP32

S = Sign bit | Exp = Exponent | Mant = Mantissa | Int = Integer | b = bits

Figure 10-1. *This figure compares numeric formats for model quantization, showing the bit allocation and compression ratios relative to FP32. FP32 (32 bits) serves as the baseline, while FP16 (16 bits) provides 50% reduction, INT8 (8 bits) achieves 75% reduction, and 4-bit formats (NF4, FP4) deliver maximum 87.5% compression. Each format allocates bits among sign, exponent, and mantissa/ integer components, with lower precision enabling smaller model sizes*

- **Reduced Memory Footprint:** Lower precision means each parameter takes up less space. An INT8 model uses half the memory of its FP16 counterpart, and an INT4 model uses only a quarter as shown in Figure 10-1. This allows larger models to fit on smaller, more accessible hardware.

- **Faster Inference:** Computations with lower-precision numbers, especially integers, can be significantly faster on compatible hardware, leading to lower latency for the end user.

The fundamental trade-off is that reducing precision inevitably involves some loss of information. The primary challenge of quantization is to compress the model as aggressively as possible while minimizing the impact on its accuracy and performance.

Post-Training Quantization (PTQ)

The most common and straightforward approach is **Post-Training Quantization (PTQ)**. This involves quantizing a model that has already been fully trained or fine-tuned. It is highly practical because it doesn't require any changes to the original training process.

The simplest way to apply PTQ is through libraries like bitsandbytes (`https://github.com/bitsandbytes-foundation/bitsandbytes`), which is seamlessly integrated into the Hugging Face ecosystem. With a few simple flags, you can load any compatible model with its weights already quantized to 8-bit or 4-bit precision. The 4-bit implementation often uses the **NF4 (NormalFloat4)** data type, a format specifically designed for the bell-curve distribution of neural network weights, which preserves accuracy better than standard 4-bit integers.

bitsandbytes makes Large Language Models (LLMs) more accessible through **k-bit quantization**. Its main goal is to dramatically reduce memory consumption for both LLM inference and training.

The library provides three main features:

- **8-bit Optimizers:** It uses a technique called block-wise quantization to maintain the performance of 32-bit optimizers while significantly reducing the memory cost.

- **LLM.int8() (8-bit Quantization):** This feature enables large language model inference with about half the required memory and is designed to avoid performance degradation. It works by quantizing most features to 8 bits using vector-wise quantization while separately handling statistical "outliers" with 16-bit matrix multiplication.

- **QLoRA (4-bit Quantization):** This method aims to make training LLMs more memory efficient. It quantizes the model to 4 bits and, crucially, inserts a small set of trainable low-rank adaptation (LoRA) weights. This allows the model to be fine-tuned without compromising performance, even on more constrained hardware.

The library achieves this by providing quantization primitives for 4-bit and 8-bit operations (e.g., `bitsandbytes.nn.Linear4bit`) and 8-bit optimizers (e.g., in the `bitsandbytes.optim` module).

Recipe 1: Effortless Quantization with bitsandbytes

Goal: Load a pre-trained model in 8-bit and 4-bit precision for efficient inference.

This procedure demonstrates **Post-Training Quantization (PTQ)** using the **bitsandbytes** library, which is integrated with the Hugging Face `transformers` ecosystem. The core objective is to reduce the Large Language Model's memory footprint and accelerate inference by converting its high-precision weights (like FP16/ BF16) to lower-precision 8-bit and 4-bit formats.

The quantization workflow for this recipe follows these general steps:

1. **Load Tokenizer and Configure:** Load the tokenizer for the chosen model (e.g., `google/gemma-2b`) and ensure proper padding/EOS tokens are set.

2. **Load Native Precision Model (Reference):** Load the model in its original precision (BF16/FP16) to establish a baseline for memory usage and performance.

3. **Load 8-bit Quantized Model:** Load the model using a `BitsAndBytesConfig` with `load_in_8bit=True` to achieve a significant memory reduction.

4. **Load 4-bit Quantized Model (NF4):** Load the model with `load_in_4bit=True` and configure it to use the memory-efficient **NF4 (NormalFloat4)** data type for maximum compression.

5. **Test Inference (Optional):** Run a brief text generation task on the quantized models to verify functionality and compare inference speed.

What to Expect

This recipe dramatically reduces a Large Language Model's memory footprint, achieving up to a 75% reduction with 4-bit NF4 quantization. This enables significantly larger models to be deployed on consumer GPUs while also delivering faster inference. This is a critical step in transforming memory-intensive models into lean, production-ready artifacts by maximizing compression while preserving accuracy through specialized post-training quantization.

```
# --- Recipe: Effortless Shrinking with `bitsandbytes` ---
```

```python
# Goal: Load a pre-trained model using 8-bit and 4-bit quantization via
transformers + bitsandbytes.
# Libraries: transformers, torch, accelerate, bitsandbytes, sentencepiece
# Note: Requires `bitsandbytes` installation. `accelerate` needed for
`device_map`.

import torch
from transformers import AutoTokenizer, AutoModelForCausalLM,
BitsAndBytesConfig
import time # For basic timing

# --- Configuration ---
MODEL_ID = "google/gemma-2b" # Choose a model
# MODEL_ID = "mistralai/Mistral-7B-v0.1" # Larger model to see more
significant memory savings

# --- 1. Load Tokenizer ---
print(f"Loading tokenizer for: {MODEL_ID}")
try:
    tokenizer = AutoTokenizer.from_pretrained(MODEL_ID)
    if tokenizer.pad_token is None:
        tokenizer.pad_token = tokenizer.eos_token
except Exception as e:
    print(f"Error loading tokenizer: {e}")
    exit()

# --- 2. Load Model in Native Precision (Reference) ---
print("\n--- Loading Model in Native Precision (BF16/FP16) ---")
# Determine compute dtype
compute_dtype = torch.bfloat16 if torch.cuda.is_available() and torch.cuda.
is_bf16_supported() else torch.float16
print(f"Using compute dtype: {compute_dtype}")
try:
    model_native = AutoModelForCausalLM.from_pretrained(
        MODEL_ID,
        torch_dtype=compute_dtype,
        device_map="auto" # Use GPU if available
    )
```

```python
    print("Native model loaded.")
    mem_footprint_native = model_native.get_memory_footprint()
    print(f"Native Model Memory Footprint: {mem_footprint_native /
    1024**3:.2f} GB")
except Exception as e:
    print(f"Error loading native model: {e}")
    model_native = None # Ensure variable exists

# --- 3. Load Model in 8-bit ---
print("\n--- Loading Model in 8-bit ---")
try:
    bnb_config_8bit = BitsAndBytesConfig(load_in_8bit=True)
    model_8bit = AutoModelForCausalLM.from_pretrained(
        MODEL_ID,
        quantization_config=bnb_config_8bit,
        device_map="auto" # device_map handles quantized models too
    )
    print("8-bit model loaded.")
    mem_footprint_8bit = model_8bit.get_memory_footprint()
    print(f"8-bit Model Memory Footprint: {mem_footprint_8bit /
    1024**3:.2f} GB")
    if model_native:
        print(f"Reduction vs Native: {(1 - mem_footprint_8bit / mem_
        footprint_native) * 100:.1f}%")
except Exception as e:
    print(f"Error loading 8-bit model: {e}")
    print("Ensure 'bitsandbytes' is installed correctly.")
    model_8bit = None

# --- 4. Load Model in 4-bit (NF4) ---
print("\n--- Loading Model in 4-bit (NF4) ---")
try:
    bnb_config_4bit = BitsAndBytesConfig(
        load_in_4bit=True,
        bnb_4bit_quant_type="nf4", # NormalFloat4 data type
        bnb_4bit_compute_dtype=compute_dtype, # Compute in bf16/fp16
```

```python
        bnb_4bit_use_double_quant=True, # Enable double quantization
    )
    model_4bit = AutoModelForCausalLM.from_pretrained(
        MODEL_ID,
        quantization_config=bnb_config_4bit,
        device_map="auto"
    )
    print("4-bit model loaded.")
    mem_footprint_4bit = model_4bit.get_memory_footprint()
    print(f"4-bit Model Memory Footprint: {mem_footprint_4bit /
    1024**3:.2f} GB")
    if model_native:
        print(f"Reduction vs Native: {(1 - mem_footprint_4bit / mem_
        footprint_native) * 100:.1f}%")
except Exception as e:
    print(f"Error loading 4-bit model: {e}")
    print("Ensure 'bitsandbytes' is installed correctly.")
    model_4bit = None

# --- 5. Test Inference (Optional) ---
# Run generation to see if models work after loading
prompt = "Instruction: Write a short description of quantization.\
nResponse:"
max_new_tokens_inf = 50

def run_inference(model, model_name):
    if model is None:
        print(f"\nSkipping inference for {model_name} (not loaded).")
        return
    print(f"\n--- Running Inference ({model_name}) ---")
    print(f"Prompt: {prompt}")
    try:
        inputs = tokenizer(prompt, return_tensors="pt").to(model.device)
        start_time = time.time()
        outputs = model.generate(
            **inputs,
            max_new_tokens=max_new_tokens_inf,
```

```
            pad_token_id=tokenizer.eos_token_id # Use EOS token ID for
                                        padding in generation
        )
        end_time = time.time()
        response = tokenizer.decode(outputs[0], skip_special_tokens=True)
        print(f"Response:\n{response}")
        print(f"Inference Time: {end_time - start_time:.2f} seconds")
    except Exception as e:
        print(f"Error during {model_name} inference: {e}")

# Run inference on loaded models
# run_inference(model_native, "Native Precision") # Can be slow
run_inference(model_8bit, "8-bit")
run_inference(model_4bit, "4-bit")

print("\nNote: Memory footprint is approximate. Inference time depends
heavily on hardware.")
# --- End of Recipe ---
```

The **bitsandbytes** approach provides the fastest and simplest path to achieving significant memory savings and inference speedup through quantization. While it is excellent for ease of use and a common first step, use cases that demand even greater control and minimal accuracy degradation at the lowest bitrates must turn to more mathematically sophisticated Post-Training Quantization methods. This leads us to **GPTQ** and **AWQ**, which offer enhanced precision at the cost of a slight increase in implementation complexity.

Generative Pre-trained Transformer Quantization (GPTQ) is an advanced Post-Training Quantization (PTQ) method designed to achieve near-lossless model compression, particularly at extremely low bit widths (3- and 4-bit). GPTQ (Generative Pre-trained Transformer Quantization) is used when minimal accuracy degradation is critical, especially at low bit rates such as 4-bit or 3-bit. While the simpler bitsandbytes method is like using a fast, basic photo editor to batch compress a high-resolution photograph to 4-bit color, saving massive space but potentially causing noticeable quality loss in complex areas, GPTQ acts like a state-of-the-art image compression codec. It achieves superior results by working layer by layer (region by region, in the analogy), where its Hessian-based error minimization functions as an intelligent "quality control eye." This process meticulously analyzes the impact of compression on one region and, when that compression introduces error, deliberately tweaks neighboring regions to cancel that error out so mistakes don't accumulate. Under the hood, this meticulous process includes identifying critical weights through activation analysis, applying per channel scaling, and optimizing zero points, requiring only a small calibration dataset (128 to 1024 samples) to achieve near-lossless 4-bit compression. This makes GPTQ a slower but intelligent and compensatory process that achieves the same small file size with dramatically superior model quality. This is why a GPTQ-quantized 4-bit Llama 70B can match the full-precision model on benchmarks, whereas a naively quantized version shows visible degradation. Figure 10-2 illustrates the detailed GPTQ quantization process, outlining the layer-by-layer optimization that transforms a high-precision model into a compressed, near-lossless INT4/INT3 format.

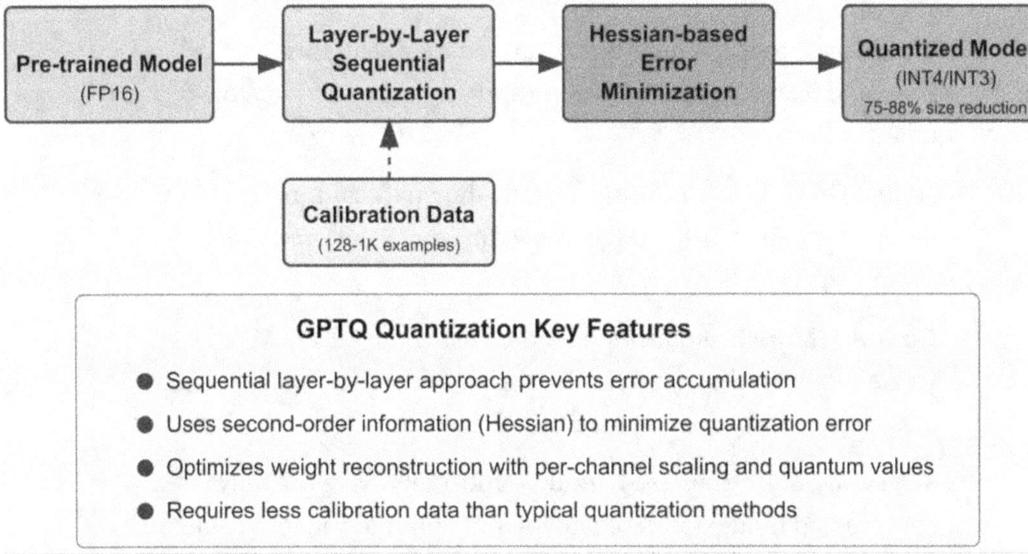

Figure 10-2. *This figure depicts the GPTQ quantization process, showing the transformation from a pre-trained FP16 model to a compressed INT4/INT3 model. The workflow includes activation sensitivity analysis using calibration data, per channel scale optimization, and final quantization with 75–88% size reduction. GPTQ's key innovation is its layer-by-layer approach using Hessian-based error minimization, which identifies critical weights through activation analysis, applies per channel scaling, optimizes zero points, and implements hardware-aware optimizations, all requiring only 128 to 1024 calibration samples to achieve near-lossless 4-bit compression*

Recipe 2: Precise Compression with GPTQ

Goal: Quantize a model using the optimum and auto-gptq libraries, which requires a calibration dataset.

This procedure demonstrates **Post-Training Quantization (PTQ)** using the **AutoGPTQ** library, which implements the **GPTQ (Generative Pre-trained Transformer Quantization)** algorithm. The core objective is to achieve highly precise compression by quantizing a Large Language Model's high-precision weights (e.g., FP16) into ultra-low-precision formats (e.g., 4-bit) through a mathematically optimized, layer-by-layer process that minimizes quantization error using a small calibration dataset.

The quantization workflow for this recipe follows these general steps:

1. **Load Tokenizer and Calibration Data:** Load the tokenizer for the chosen model and prepare a small, representative calibration dataset (e.g., 128–1K samples), which is essential for quantization.

2. **Load Base Model:** Load the pre-trained model at its native precision to establish the baseline and serve as the target for quantization.

3. **Define Quantization Config:** Specify the quantization parameters, such as the target bit width (e.g., 4-bit) and group size, using `BaseQuantizeConfig`.

4. **Quantize Model:** Quantize the model using the `AutoGPTQFor CausalLM.quantize()` method, which applies layer-wise Hessian-based optimization using the calibration data to minimize error.

5. **Save and Load Quantized Model (Optional):** Save the fully compressed model for deployment and optionally load it back to verify functionality and performance.

What to Expect

This recipe is designed for precise compression, achieving an estimated 75% to 88% reduction in the model's memory footprint, which enables deployment on smaller, more accessible GPUs. GPTQ's core innovation is layer-wise optimization using a small calibration set that is designed to deliver **Very Good** quality retention and near-lossless compression at 4 bits. The main trade-off is the process complexity and time investment, which can take approximately 1 to 3 hours to complete.

```python
import torch
from transformers import AutoTokenizer, AutoModelForCausalLM,
TextGenerationPipeline
from datasets import load_dataset
from auto_gptq import AutoGPTQForCausalLM, BaseQuantizeConfig # AutoGPTQ
specific imports
import time
import logging
# --- Configuration -
# Choose a base model supported by AutoGPTQ (check their GitHub)
# Smaller models quantize faster. gpt2 variants are common examples.
MODEL_CHECKPOINT = "gpt2-medium" # ~355M parameters
# Calibration dataset - needs to be representative of text the model
will see
CALIBRATION_DATASET = "allenai/c4"
CALIBRATION_SPLIT = "train" # Use train split
NUM_CALIBRATION_SAMPLES = 128 # Number of samples for calibration
(e.g., 128)
CALIBRATION_SEQ_LEN = 512 # Sequence length for calibration data
# GPTQ Quantization Config
QUANTIZE_BITS = 4 # Target bits (e.g., 4, 3, 8)
QUANTIZE_GROUP_SIZE = 128 # Group size for quantization (e.g., 32, 64, 128,
-1 for per-channel)
QUANTIZE_DESC_ACT = False # Or True - Whether to quantize using act_
order=True, can improve accuracy but slower
# Output directory for quantized model
QUANTIZED_MODEL_DIR = f"./{MODEL_CHECKPOINT.split('/')[-1]}-gptq-{QUANTIZE_
BITS}bit"

# Setup logging for AutoGPTQ
logging.basicConfig(
    format="%(asctime)s %(levelname)s [%(name)s] %(message)s",
level=logging.INFO, datefmt="%Y-%m-%d %H:%M:%S"
)
```

```python
# --- 1. Load Tokenizer and Calibration Data ---
print(f"Loading tokenizer: {MODEL_CHECKPOINT}")
try:
    tokenizer = AutoTokenizer.from_pretrained(MODEL_CHECKPOINT)
    if tokenizer.pad_token is None: tokenizer.pad_token = tokenizer.
    eos_token
except Exception as e: print(f"Error loading tokenizer: {e}"); exit()

print(f"\nLoading calibration data: {CALIBRATION_DATASET} (subset)")
try:
    # Load calibration data (streaming recommended for large datasets
    like C4)
    calibration_dataset = load_dataset(CALIBRATION_DATASET, name="en",
    split=CALIBRATION_SPLIT, streaming=True)
    # Take a sample and tokenize
    samples = []
    for data in calibration_dataset.take(NUM_CALIBRATION_SAMPLES):
        # Tokenize, ensuring padding/truncation to fixed length for
        calibration
        tokenized_sample = tokenizer(data['text'], return_tensors='pt',
        max_length=CALIBRATION_SEQ_LEN, padding='max_length',
        truncation=True)
        samples.append({
        "input_ids": tokenized_sample["input_ids"].squeeze(0),
        # LongTensor [SEQ]
        "attention_mask": tokenized_sample["attention_mask"].squeeze(0)
        # LongTensor [SEQ]
        })
        # Alternative: provide list of strings directly to quantize method
        if supported by backend
        # samples.append(data['text'])
    if not samples: raise ValueError("No calibration samples loaded.")
    print(f"Loaded {len(samples)} calibration samples.")
    # If using input_ids, stack them if needed by quantize method,
    otherwise keep as list
    # calibration_data_final = torch.stack(samples)
```

```python
    # AutoGPTQ often expects a list of strings or dicts
    #calibration_data_final = [tokenizer.decode(s, skip_special_
    tokens=True) for s in samples]
    calibration_data_final = samples

except Exception as e:
    print(f"Error loading or processing calibration data: {e}")
    exit()

# --- 2. Load Base Model ---
print(f"\nLoading base model: {MODEL_CHECKPOINT}")
try:
    # Load in native precision on CPU first maybe, or directly to GPU if
    memory allows
    model = AutoModelForCausalLM.from_pretrained(MODEL_CHECKPOINT, torch_
    dtype=torch.float16, low_cpu_mem_usage=True)
    # model.to('cuda:0') # Move to GPU if not done automatically
    print("Base model loaded.")
except Exception as e:
    print(f"Error loading base model: {e}")
    exit()

# --- 3. Define Quantization Config ---
print("\nDefining GPTQ quantization config...")
quantize_config = BaseQuantizeConfig(
    bits=QUANTIZE_BITS, # Number of bits for quantization
    group_size=QUANTIZE_GROUP_SIZE, # Group size
    desc_act=QUANTIZE_DESC_ACT, # Activation order; True might improve
    accuracy, False is faster
    damp_percent=0.01, # Dampening percentage for Hessian computation
    sym=True # Use symmetric quantization
)

# --- 4. Quantize Model ---
print("\nStarting GPTQ quantization process...")
print(f"Bits: {QUANTIZE_BITS}, Group Size: {QUANTIZE_GROUP_SIZE}, Desc Act:
{QUANTIZE_DESC_ACT}")
```

```python
print("This can take a while...")
start_time = time.time()
try:
    # Wrap model with AutoGPTQ wrapper
    quantized_model_gptq = AutoGPTQForCausalLM.from_pretrained(
        MODEL_CHECKPOINT,
        quantize_config=quantize_config, # Pass the config
        # Optional: Pass model directly if already loaded
        # model=model, # Pass the pre-loaded model object
        torch_dtype=torch.float16, # Ensure consistency
        trust_remote_code=True, # Often needed
        device_map="auto" # Let AutoGPTQ handle device placement
    )

    # Run the quantization process
    quantized_model_gptq.quantize(
        calibration_data_final, # Pass the prepared calibration data
        batch_size=1, # Calibration batch size
        use_triton=torch.cuda.is_available(), # Use Triton kernels if
        available (faster)
        # cache_examples_on_gpu=True # If VRAM allows
    )
    end_time = time.time()
    print(f"Quantization finished in {end_time - start_time:.2f} seconds.")

    # --- 5. Save Quantized Model ---
    print(f"\nSaving quantized model to: {QUANTIZED_MODEL_DIR}")
    # Use export_quantized=True argument or specific save methods depending
    on AutoGPTQ version
    # Option 1: Standard save_pretrained (might work for newer versions)
    #quantized_model_gptq.save_pretrained(QUANTIZED_MODEL_DIR, safe_
    serialization=True)

    # Option 2: Use export_quantized (check AutoGPTQ docs for current best
    practice)
    # Example for older versions might differ
```

```python
    quantized_model_gptq.save_pretrained(QUANTIZED_MODEL_DIR, use_
    safetensors=True)

    tokenizer.save_pretrained(QUANTIZED_MODEL_DIR) # Save tokenizer too
    print("Quantized model and tokenizer saved.")

except Exception as e:
    print(f"Error during GPTQ quantization or saving: {e}")
    print("Ensure AutoGPTQ and its dependencies (like optimum) are
    installed.")
    exit()

# --- 6. Load and Test Quantized Model (Optional) ---
print("\nLoading and testing quantized model...")
try:
    # Load the quantized model using the AutoGPTQ class
    # Important: Ensure the environment loading the model has AutoGPTQ
    installed
    model_loaded_gptq = AutoGPTQForCausalLM.from_quantized(
        QUANTIZED_MODEL_DIR,
        device_map="auto", # Load onto GPU
        use_triton=torch.cuda.is_available(),
        trust_remote_code=True,
        # inject_fused_attention=True, # Optional: Speed up inference
        # inject_fused_mlp=True # Optional: Speed up inference
    )
    print("Quantized model loaded successfully.")

    # Test inference
    prompt = "Quantization in deep learning is"
    print(f"Prompt: {prompt}")
    # Use pipeline for easy generation
    #pipeline_gptq = TextGenerationPipeline(model=model_loaded_gptq,
    tokenizer=tokenizer, device=model_loaded_gptq.device)
    pipeline_gptq = TextGenerationPipeline(model=model_loaded_gptq,
    tokenizer=tokenizer)
    start_time = time.time()
```

```python
    outputs = pipeline_gptq(prompt, max_new_tokens=50, do_sample=True,
    temperature=0.7)
    end_time = time.time()
    print(f"Generated Text:\n{outputs[0]['generated_text']}")
    print(f"Inference Time: {end_time - start_time:.2f} seconds")

except Exception as e:
    print(f"Error loading or testing quantized model: {e}")
# --- End of Recipe ---
```

A good **calibration dataset** is fundamental to advanced Post-Training Quantization (PTQ) methods such as GPTQ and AWQ. This dataset is a small, carefully selected sample (typically 128 to 1024 samples) used by the quantization algorithm to analyze the model's weight and activation distributions and precisely minimize the quantization error. The most critical selection criterion is that the dataset must be representative: it should contain text that is similar in domain, style, and length to the data the model will encounter in its final production environment. Using an irrelevant calibration set will lead to poor compensation during quantization and significant accuracy degradation on real-world inputs.

Activation-Aware Weight Quantization (AWQ)

Another advanced PTQ method is **Activation-Aware Weight Quantization (AWQ)**. This technique recognizes that not all weights in a model are equally important. It identifies a tiny fraction of weights (often less than 1%) that are most critical for performance by analyzing the scale of the activations that pass through them. AWQ preserves these "salient" weights in higher precision while aggressively quantizing the rest. This hybrid approach aims to maintain high accuracy while achieving significant compression and inference speedup.

AWQ is an intelligent model compression technique that achieves a 75% size reduction while maintaining 98–99% of the model's quality through a systematic seven-step process. The journey begins with a pre-trained model in FP16 format (e.g., Llama-2–7B at 14GB) that must be compressed to approximately 4-bit precision (~3.5GB) for deployment on consumer GPUs (Figures 10-3 and 10-4).

AWQ: Activation-aware Weight Quantization

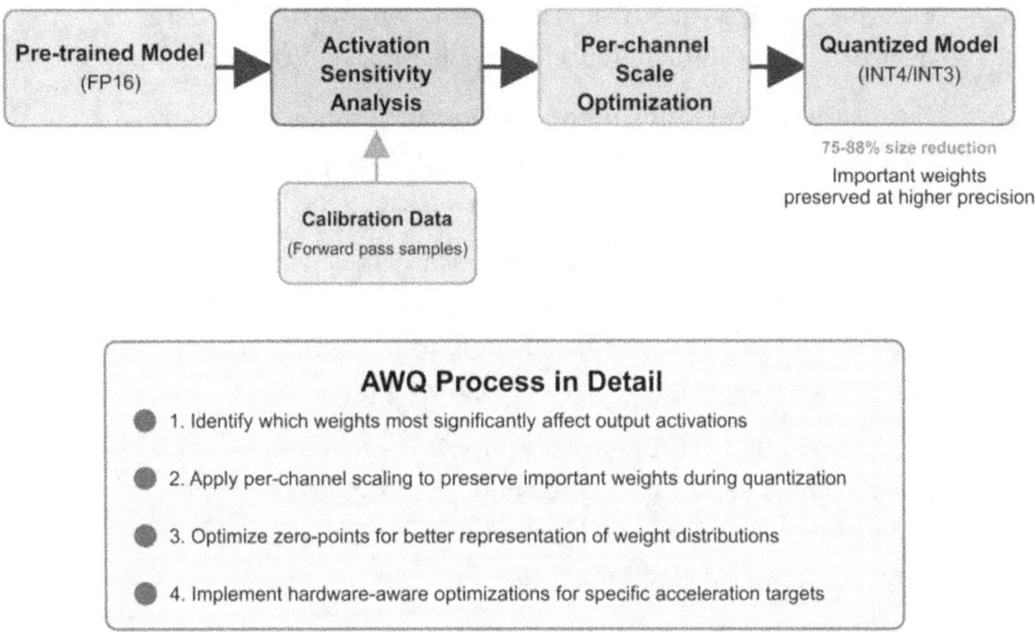

Figure 10-3. *Activation-Aware Weight Quantization: Step-by-Step Process*

AWQ compresses a pre-trained model through seven systematic steps:

1. **Calibration:** Runs a small dataset (128 to 1024 samples) through the model, observing how it processes real data without making any changes.

2. **Activation Sensitivity Measurement:** Categorizes each weight into three groups: critical (10%, such as attention Q and K matrices), moderate (30%), and low importance (60%), based on how much each weight affects the outputs.

3. **Per Channel Scaling:** Assigns protection levels based on importance. Critical weights get larger quantization buckets to preserve precision, while low-importance weights get smaller buckets for aggressive compression, like an insurance policy where expensive items get full coverage and cheap items get minimal protection.

4. **Zero Point Optimization:** Shifts where "zero" falls in the
    quantization range to match actual weight distributions,
    ensuring no precision is wasted on unused ranges, like focusing a
    microscope directly on the specimen rather than empty space.

5. **Hardware-Aware Optimization:** Tailors quantization settings to
    the deployment target, whether NVIDIA GPUs, Apple Silicon, or
    edge devices, maximizing performance on each specific platform.

6. **Mixed Precision Quantization:** Applies all gathered intelligence,
    keeping critical weights at INT8 or FP16, moderate weights at
    INT5 or INT6, and compressing low-importance weights to INT4
    or INT3, resulting in an average of roughly 4 bits per weight.

7. **Deployment:** Delivers dramatic results: 75% size reduction, 2.75x
    faster inference, 98.4% quality retention, and 75% lower serving
    costs, enabling consumer GPUs to run models that previously
    required enterprise hardware.

The magic of AWQ lies in its intelligent weight distribution: 10% of critical weights (attention mechanisms, embeddings, and output heads) are preserved at high precision, 30% of moderate weights receive medium compression, and 60% of low-importance weights undergo aggressive compression. This achieves the same 75% size reduction as naive uniform quantization, but with only 1% to 2% quality loss, compared to 5% to 10% degradation from simple approaches, all in just 30 minutes to 2 hours, with no retraining required.

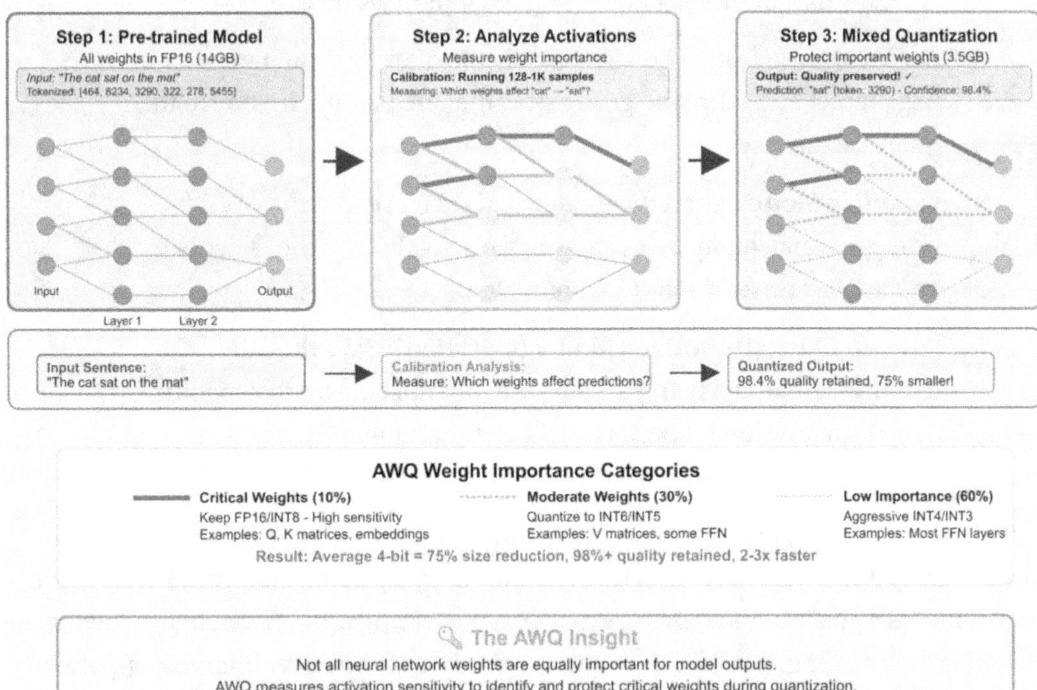

Figure 10-4. Intuitive Depiction of Three Important Steps in the AWQ Process

Recipe 3: Activation-Aware Weight Quantization (AWQ)

Goal: Quantize a model using the AWQ method, which preserves important weights to maintain accuracy.

This procedure demonstrates **Post-Training Quantization (PTQ)** using the **AutoAWQ** library, which implements the **Activation-Aware Weight Quantization (AWQ)** method. The core objective is to achieve maximum compression by preserving the most critical weights at a higher precision while aggressively quantizing the rest, thus minimizing accuracy degradation for efficient deployment.

The quantization workflow for this recipe follows these general steps:

1. **Load Tokenizer and Calibration Data:** Load the tokenizer for the chosen model and prepare a small, representative calibration dataset for the activation analysis.

2. **Load Base Model:** Prepare to load the base model, which AWQ typically handles internally during the quantization process.

3. **Define Quantization Config:** Specify the quantization parameters (e.g., target bit width and group size) that are passed directly to the quantize method.

4. **Quantize Model:** Load the model, and then execute the AWQ process, which analyzes the activation sensitivity of the weights and applies mixed precision quantization.

5. **Save and Load Quantized Model (Optional):** Save the fully compressed model using model.save_quantized() for fast loading, and optionally load it back to verify functionality.

What to Expect

This recipe is designed for **Activation-Aware Compression**, achieving approximately a 75% size reduction while retaining exceptional quality (98–99% of the original model's performance). AWQ's focus on preserving critical weights enables it to be significantly faster than GPTQ (30 minutes to 2 hours) while maintaining superior accuracy compared to basic bitsandbytes, making it an excellent choice for quickly producing lean, high-performing models for deployment.

```
# --- Recipe 3: Activation-Aware Quantization (AWQ) ---
# Goal: Quantize a pre-trained model using the Activation-aware Weight
Quantization (AWQ) method.
# Libraries: transformers, torch, optimum, datasets, autoawq
# Note: Requires installing AutoAWQ: pip install autoawq
#       Requires a GPU and is often faster than GPTQ for the same
model size.
#       Uses gpt2-medium as an example and C4 dataset for calibration.

import torch
from transformers import AutoTokenizer, AutoModelForCausalLM
from awq import AutoAWQForCausalLM
from datasets import load_dataset
import time
import logging

# --- Configuration ---
```

```python
# Choose a base model. AWQ works well with many modern architectures.
MODEL_CHECKPOINT = "gpt2-medium" # ~355M parameters
# Calibration dataset - a small, representative sample of text.
CALIBRATION_DATASET = "allenai/c4"
CALIBRATION_SPLIT = "train"
NUM_CALIBRATION_SAMPLES = 128
# AWQ Quantization Config
QUANTIZE_BITS = 4
QUANTIZE_GROUP_SIZE = 128
# Output directory for the quantized model
QUANTIZED_MODEL_DIR = f"./{MODEL_CHECKPOINT.split('/')[-1]}-awq-{QUANTIZE_
BITS}bit"

# Setup logging
logging.basicConfig(
    format="%(asctime)s %(levelname)s [%(name)s] %(message)s",
level=logging.INFO, datefmt="%Y-%m-%d %H:%M:%S"
)

# --- 1. Load Tokenizer and Calibration Data ---
print(f"Loading tokenizer: {MODEL_CHECKPOINT}")
try:
    tokenizer = AutoTokenizer.from_pretrained(MODEL_CHECKPOINT)
    if tokenizer.pad_token is None: tokenizer.pad_token = tokenizer.
    eos_token
except Exception as e:
    print(f"Error loading tokenizer: {e}"); exit()

print(f"\nLoading calibration data: {CALIBRATION_DATASET} (subset)")
try:
    # Load a small subset of the data for calibration
    calibration_dataset = load_dataset(CALIBRATION_DATASET, name="en",
    split=f"{CALIBRATION_SPLIT}[:{NUM_CALIBRATION_SAMPLES}]")
    # AWQ expects a list of strings
    calibration_data_final = [example['text'] for example in calibration_
    dataset]
    print(f"Loaded {len(calibration_data_final)} calibration samples.")
```

```python
except Exception as e:
    print(f"Error loading or processing calibration data: {e}")
    exit()

# --- 2. Load Base Model ---
# AWQ handles loading the model internally, so we just need the path.
print(f"\nPreparing to load base model for AWQ: {MODEL_CHECKPOINT}")

# --- 3. Define Quantization Config ---
# For AWQ, the configuration is passed directly to the quantize method.
# The core parameters are the number of bits and the group size.
awq_config = {
    "w_bit": QUANTIZE_BITS,
    "q_group_size": QUANTIZE_GROUP_SIZE,
    "zero_point": True # Use a zero point for better accuracy
}
print(f"\nDefined AWQ config: {awq_config}")

# --- 4. Quantize Model ---
print("\nStarting AWQ quantization process...")
print("This involves loading the model, analyzing activations, and
quantizing weights.")
start_time = time.time()
try:
    # Load the model and quantize it in one step
    model = AutoAWQForCausalLM.from_pretrained(MODEL_CHECKPOINT, low_cpu_
    mem_usage=True, device_map="auto")

    # Run the quantization process
    model.quantize(
        tokenizer,
        quant_config=awq_config,
        calo_data=calibration_data_final
    )
    end_time = time.time()
    print(f"Quantization finished in {end_time - start_time:.2f} seconds.")

    # --- 5. Save Quantized Model ---
```

```python
    print(f"\nSaving quantized model to: {QUANTIZED_MODEL_DIR}")
    # The `save_quantized` method saves the model in a format that can be
    loaded for fast inference.
    model.save_quantized(QUANTIZED_MODEL_DIR)
    tokenizer.save_pretrained(QUANTIZED_MODEL_DIR)
    print("Quantized model and tokenizer saved.")

except Exception as e:
    print(f"Error during AWQ quantization or saving: {e}")
    print("Ensure AutoAWQ is installed correctly.")
    exit()

# --- 6. Load and Test Quantized Model (Optional) ---
print("\nLoading and testing quantized model...")
try:
    # Load the quantized model using the AutoAWQ class again
    model_quantized = AutoAWQForCausalLM.from_quantized(QUANTIZED_MODEL_
    DIR, device_map="auto")
    print("Quantized model loaded successfully.")

    # Test inference
    prompt = "Activation-aware Weight Quantization is a technique that"
    print(f"Prompt: {prompt}")

    # Use the model directly for generation
    tokens = tokenizer(prompt, return_tensors="pt").to("cuda")
    start_time = time.time()
    outputs = model_quantized.generate(**tokens, max_new_tokens=50, do_
    sample=True, temperature=0.7)
    end_time = time.time()

    generated_text = tokenizer.decode(outputs[0], skip_special_tokens=True)
    print(f"Generated Text:\n{generated_text}")
    print(f"Inference Time: {end_time - start_time:.2f} seconds")

except Exception as e:
    print(f"Error loading or testing quantized model: {e}")

# --- End of Recipe ---
```

Having covered the conceptual details of Post-Training Quantization techniques (bitsandbytes, GPTQ, and AWQ), along with the alternative of QAT, you now have a comprehensive view of the available methods. To make an informed deployment decision, the critical final step is to empirically analyze the trade-offs in performance, cost, and complexity between these techniques, as summarized in Table 10-1.

Table 10-1. *Comparison of QAT, GPTQ, and AWQ Methods*

Aspect	QAT (Quantization-Aware Training)	AWQ (Activation-Aware Weight Quantization)	GPTQ (Generative Pre-trained Transformer Quantization)
Timing	During training process	After training (Post-Training)	After training (Post-Training)
Model Weights	Modified through training	Preserved, only precision changes	Preserved, only precision changes
Computational Cost	Very High—Requires full training	Moderate—Analysis + quantization	Moderate to High—Layer-by-layer optimization
Data Requirement	Full training dataset	Small calibration set (128–1K samples)	Small calibration set (128–1K samples)
Time Investment	Days to weeks	30 minutes–2 hours	1–3 hours
Quality	Best results—Model trained for quantization	Excellent—Preserves critical weights	Very Good—Mathematically optimal
Approach	Simulates quantization during training	Identifies activation-sensitive weights	Layer-wise Hessian-based optimization
Hardware Awareness	Can be incorporated during training	Strong—Optimized for target hardware	Limited hardware-specific optimizations
Mathematical Foundation	Gradient-based learning	Activation magnitude analysis	Second-order Hessian optimization

(continued)

Table 10-1. (*continued*)

Aspect	QAT (Quantization-Aware Training)	AWQ (Activation-Aware Weight Quantization)	GPTQ (Generative Pre-trained Transformer Quantization)
Implementation Complexity	High—Requires training infrastructure	Medium—Analysis algorithms	Medium—Mathematical optimization
GPU Memory During Process	High—Full model + gradients + optimizer	Medium—Model + activation analysis	Medium—Model + layer-wise processing
Bit width Support	1–8 bits (excellent at low bits)	Typically 3–4 bits optimal	Typically 3–4 bits optimal

Quantization-Aware Training (QAT)

An alternative to PTQ is **Quantization-Aware Training (QAT)**. Instead of quantizing a fully trained model, QAT simulates the effects of quantization *during* fine-tuning. It inserts "fake quantization" operations into the model graph, allowing the model's weights to adapt and compensate for precision loss during training.

While QAT can potentially achieve higher accuracy than PTQ, especially for very low bitrates, it is more complex to implement, requires modifications to the training loop, and increases training time. We are not implementing QAT in this chapter. For most common fine-tuning workflows, combining PTQ on the base model with PEFT techniques like LoRA (as seen in the QLoRA method) offers a more practical and efficient path to creating specialized, lightweight models.

Comparative Analysis of Quantization Methods

The choice between QAT, AWQ, and GPTQ depends on a trade-off between performance, computational cost, and implementation complexity. Table 10-1 provides a detailed comparison of these three prominent techniques.

Measuring Quantization Performance

To make an informed decision about which quantization method to use, it's essential to measure the trade-offs empirically. The two most important metrics are the reduction in memory footprint and the change in inference latency (speed).

Recipe 4: Measuring Quantization Gains

Goal: Compare the memory usage and inference speed of a model before and after quantization.

This procedure demonstrates the empirical measurement of quantization benefits using the **bitsandbytes** 4-bit (NF4) method. The core objective is to quantify the trade-offs of quantization by comparing the performance and memory usage of a model in its native high-precision format against its compressed, 4-bit-quantized version in a single script.

The performance measurement workflow for this recipe follows these general steps:

1. **Load Tokenizer:** Initialize the tokenizer for the chosen model (e.g., google/gemma-2b).

2. **Load Native Model and Measure:** Load the model in its native precision (BF16/FP16), empirically measure its GPU Memory Footprint, and calculate its Average Inference Latency using a series of warmup and timed generation runs.

3. **Load 4-bit Quantized Model and Measure:** Load the model in 4-bit (NF4) precision using BitsAndBytesConfig, measure its GPU Memory Footprint, and calculate its Average Inference Latency.

4. **Comparison Summary:** Output a final summary comparing the Native and 4-bit results for memory and speed metrics.

What to Expect

This recipe is designed to empirically show the quantization gains. You should expect a dramatic reduction in the model's static memory footprint (up to 75% with 4-bit quantization) and improved inference speed (lower latency), demonstrating the efficiency benefits of running the quantized model on GPU hardware.

```python
# --- Recipe 4: Measuring the Gains (Quantization Performance) ---
# Goal: Compare memory usage and inference speed before and after
quantization.
# Method: Uses BitsAndBytes 4-bit quantization for easy comparison within
one script.
# Libraries: transformers, torch, accelerate, bitsandbytes,
sentencepiece, time

import torch
from transformers import AutoTokenizer, AutoModelForCausalLM,
BitsAndBytesConfig
import time
import numpy as np

# --- Configuration ---
MODEL_ID = "google/gemma-2b" # Choose a model to test
# MODEL_ID = "gpt2-large" # Another option
PROMPT = "Explain the concept of transfer learning in machine learning in
about 50 words."
NUM_TOKENS_TO_GENERATE = 100
NUM_INFERENCE_RUNS = 5 # Number of times to run inference for
averaging speed

# --- 1. Load Tokenizer ---
print(f"Loading tokenizer for: {MODEL_ID}")
try:
    tokenizer = AutoTokenizer.from_pretrained(MODEL_ID)
    if tokenizer.pad_token is None: tokenizer.pad_token = tokenizer.
    eos_token
except Exception as e: print(f"Error loading tokenizer: {e}"); exit()

# --- 2. Load Native Model & Measure ---
print("\n--- Loading and Measuring Native Model (BF16/FP16) ---")
native_results = {"memory_gb": "N/A", "avg_latency_s": "N/A"}
model_native = None # Define variable outside try block
try:
    compute_dtype = torch.bfloat16 if torch.cuda.is_available() and torch.
    cuda.is_bf16_supported() else torch.float16
```

```python
model_native = AutoModelForCausalLM.from_pretrained(
    MODEL_ID,
    torch_dtype=compute_dtype,
    device_map="auto"
)
if model_native.config.pad_token_id is None: model_native.config.pad_
token_id = tokenizer.pad_token_id
print(f"Native model loaded in {compute_dtype}.")

# Measure Memory
mem_footprint_native = model_native.get_memory_footprint()
native_results["memory_gb"] = mem_footprint_native / 1024**3
print(f"Native Memory Footprint: {native_results['memory_gb']:.2f} GB")

# Measure Inference Speed
print(f"Running inference ({NUM_INFERENCE_RUNS} runs)...")
latencies = []
inputs = tokenizer(PROMPT, return_tensors="pt").to(model_native.device)
for _ in range(NUM_INFERENCE_RUNS + 1): # +1 for warmup run
    torch.cuda.synchronize() # Ensure sync before timing
    start_time = time.time()
    _ = model_native.generate(
        **inputs,
        max_new_tokens=NUM_TOKENS_TO_GENERATE,
        pad_token_id=tokenizer.eos_token_id,
        do_sample=False # Use greedy for consistent timing
    )
    torch.cuda.synchronize() # Ensure sync after generation
    end_time = time.time()
    latencies.append(end_time - start_time)

avg_latency = np.mean(latencies[1:]) # Exclude warmup run
native_results["avg_latency_s"] = avg_latency
print(f"Native Avg. Latency ({NUM_TOKENS_TO_GENERATE} tokens): {avg_
latency:.3f} seconds")

# Clean up memory
```

```python
    del model_native
    torch.cuda.empty_cache()
    print("Native model unloaded.")

except Exception as e:
    print(f"Error during native model loading or inference: {e}")
    if 'model_native' in locals() and model_native is not None: del
    model_native
    torch.cuda.empty_cache()

# --- 3. Load 4-bit Quantized Model & Measure ---
print("\n--- Loading and Measuring 4-bit Model (NF4) ---")
quantized_results = {"memory_gb": "N/A", "avg_latency_s": "N/A"}
model_4bit = None # Define variable outside try block
try:
    compute_dtype = torch.bfloat16 if torch.cuda.is_available() and torch.
    cuda.is_bf16_supported() else torch.float16
    bnb_config_4bit = BitsAndBytesConfig(
        load_in_4bit=True,
        bnb_4bit_quant_type="nf4",
        bnb_4bit_compute_dtype=compute_dtype,
        bnb_4bit_use_double_quant=True,
    )
    model_4bit = AutoModelForCausalLM.from_pretrained(
        MODEL_ID,
        quantization_config=bnb_config_4bit,
        device_map="auto"
    )
    if model_4bit.config.pad_token_id is None: model_4bit.config.pad_token_
    id = tokenizer.pad_token_id
    print("4-bit model loaded.")

    # Measure Memory
    mem_footprint_4bit = model_4bit.get_memory_footprint()
    quantized_results["memory_gb"] = mem_footprint_4bit / 1024**3
    print(f"4-bit Memory Footprint: {quantized_results['memory_
    gb']:.2f} GB")
```

```python
    # Measure Inference Speed
    print(f"Running inference ({NUM_INFERENCE_RUNS} runs)...")
    latencies_4bit = []
    inputs = tokenizer(PROMPT, return_tensors="pt").to(model_4bit.device)
    for _ in range(NUM_INFERENCE_RUNS + 1): # Warmup run
        torch.cuda.synchronize()
        start_time = time.time()
        _ = model_4bit.generate(
            **inputs,
            max_new_tokens=NUM_TOKENS_TO_GENERATE,
            pad_token_id=tokenizer.eos_token_id,
            do_sample=False
        )
        torch.cuda.synchronize()
        end_time = time.time()
        latencies_4bit.append(end_time - start_time)

    avg_latency_4bit = np.mean(latencies_4bit[1:]) # Exclude warmup
    quantized_results["avg_latency_s"] = avg_latency_4bit
    print(f"4-bit Avg. Latency ({NUM_TOKENS_TO_GENERATE} tokens): {avg_
    latency_4bit:.3f} seconds")

    # Clean up memory
    del model_4bit
    torch.cuda.empty_cache()
    print("4-bit model unloaded.")

except Exception as e:
    print(f"Error during 4-bit model loading or inference: {e}")
    if 'model_4bit' in locals() and model_4bit is not None: del model_4bit
    torch.cuda.empty_cache()

# --- 4. Comparison Summary ---
print("\n--- Performance Comparison Summary ---")
dtype_str = str(compute_dtype).replace("torch.", "") if isinstance(compute_
dtype, torch.dtype) else "N/A"
```

```python
print(f"Metric                          | Native ({dtype_str}) | 4-bit (NF4)")
print(f"------------------------|-------------------|----------------")
mem_native_str = f"{native_results['memory_gb']:.2f} GB" if
isinstance(native_results['memory_gb'], float) else native_
results['memory_gb']
mem_4bit_str = f"{quantized_results['memory_gb']:.2f} GB" if
isinstance(quantized_results['memory_gb'], float) else quantized_
results['memory_gb']
print(f"Memory Footprint        | {mem_native_str:<17} | {mem_4bit_str:<15}")

lat_native_str = f"{native_results['avg_latency_s']:.3f} s" if
isinstance(native_results['avg_latency_s'], float) else native_
results['avg_latency_s']
lat_4bit_str = f"{quantized_results['avg_latency_s']:.3f} s" if
isinstance(quantized_results['avg_latency_s'], float) else quantized_
results['avg_latency_s']
print(f"Avg. Latency ({NUM_TOKENS_TO_GENERATE} toks) | {lat_native_str:<17}
| {lat_4bit_str:<15}")

# Calculate relative changes if possible
if isinstance(native_results['memory_gb'], float) and isinstance(quantized_
results['memory_gb'], float):
    mem_reduction = (1 - quantized_results['memory_gb'] / native_
    results['memory_gb']) * 100
    print(f"\nMemory Reduction (4-bit vs Native): {mem_reduction:.1f}%")
if isinstance(native_results['avg_latency_s'], float) and
isinstance(quantized_results['avg_latency_s'], float):
    speedup = native_results['avg_latency_s'] / quantized_results['avg_
    latency_s']
    print(f"Inference Speedup (4-bit vs Native): {speedup:.2f}x")

print("\nNote: Results are indicative and highly dependent on hardware,
model, batch size, and specific generation parameters.")

# --- End of Recipe ---
```

Memory Footprint Note: The reported memory footprint primarily reflects the size of the quantized model weights. Actual VRAM usage during inference will be higher, as it includes dynamic components like the KV cache (which grows with context length) and activation memory.

It is essential to note that quantization is not a universal solution. It should be approached with extreme caution in ultra-high-stakes applications (e.g., certain medical systems or autonomous control software) where *any* minimal, non-compensable accuracy loss, even if only 1–2%, is unacceptable and could lead to severe consequences. Furthermore, the success of quantization is tied to the hardware: older or specialized architectures may be incompatible, preventing efficient low-precision integer compute units (e.g., INT4 or INT8 support) and potentially resulting in slower performance.

Chapter Summary

In this chapter, you learned that model quantization is the critical final optimization stage for deploying computationally intensive Large Language Models (LLMs). This process addresses the significant memory and inference speed challenges by converting the model's high-precision weights (FP32/FP16) to lower-precision formats (INT8/INT4), thereby achieving a dramatic reduction in memory footprint (up to 75%) and faster inference. You explored three major Post-Training Quantization (PTQ) techniques, such as bitsandbytes, GPTQ, and AWQ, as well as the more complex alternative, Quantization-Aware Training (QAT), which is summarized for comparative analysis.

The practical knowledge was delivered through four code implementations: you started with the simple, fast bitsandbytes approach and then moved to the mathematically precise, layer-by-layer GPTQ method, and the activation-aware, high-quality AWQ process. The final and most important step was covered in Recipe 4, which showed you how to empirically measure the resulting performance and memory gains, allowing you to use the model's memory footprint and average inference latency to make an informed, data-driven decision on which quantization strategy is best for your specific hardware and deployment needs. With the model now optimized, the next chapter will focus on the final steps of bringing it to production, covering deployment and inference strategies.

Inference and Deployment Strategies

In Chapter 10, we optimized our model through quantization, reducing its memory footprint and computational requirements. However, a model artifact residing on disk does not generate value. To operationalize the model, we must bridge the gap between static weights and a dynamic, scalable service.

This chapter focuses on Inference Serving and Deployment. We will transition from simple execution scripts to production-grade serving architectures. We will explore advanced optimization techniques such as Speculative Decoding and Tensor Parallelism, compare high-throughput engines such as TGI and vLLM, and implement robust deployment patterns using Docker.

The Deployment Challenge: Latency vs. Throughput

In production environments, we optimize for two conflicting metrics:

1. **Latency:** The time taken to generate the first token (Time to First Token (TTFT)) and subsequent tokens (Inter-Token Latency). Low latency is critical for user experience in chat applications.

2. **Throughput:** The total number of requests or tokens the system can process per second. High throughput is essential for cost-efficiency and handling concurrent traffic.

Standard, unoptimized Python scripts (like those used in training) suffer from **static batching** and poor memory management, making them unsuitable for production traffic.

Advanced Inference Optimizations

To address these challenges, modern serving engines employ several sophisticated techniques.

1. Continuous Batching

Traditional batching waits for a set of requests to complete before starting new ones. If one request generates 100 tokens and another generates 10, the GPU sits idle while the shorter request waits for the longer one to finish.

Continuous Batching (or cellular batching) operates at the iteration level. As soon as a request finishes generation, it is ejected from the batch, and a new request from the queue is immediately scheduled in its place. This ensures near 100% GPU utilization. Figure 11-1 illustrates the difference between static and continuous batching.

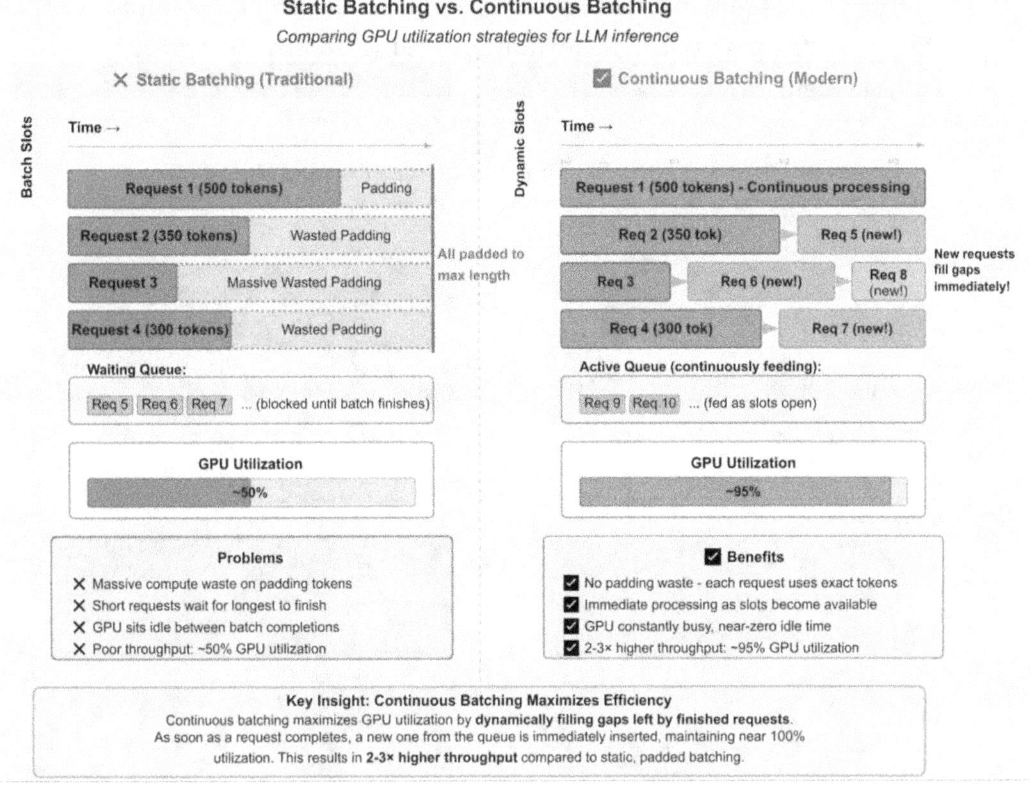

Figure 11-1. *Static vs. Continuous Batching. Traditional static batching pads all requests onto the longest sequence (500 tokens), wasting 50% of GPU compute. Continuous batching dynamically fills slots as requests complete. When request two finishes, request five immediately starts. Achieves 95% GPU utilization (vs. 50%), delivering 2–3× higher throughput*

2. Speculative Decoding

Generating text is memory bound; the GPU spends more time moving data (weights) than computing. Speculative Decoding leverages the below strategy to overcome memory constraint (Figure 11-2).

- **Drafting:** A small, fast "draft model" (e.g., a 125M parameter model) generates a sequence of 5–10 candidate tokens very quickly.

- **Verification:** The large "target model" (e.g., 70B parameter) processes all draft tokens in a single parallel forward pass to verify them.

- **Result:** If the draft tokens are correct, we effectively generate multiple tokens at the cost of a single forward pass. If incorrect, we discard and correct them.

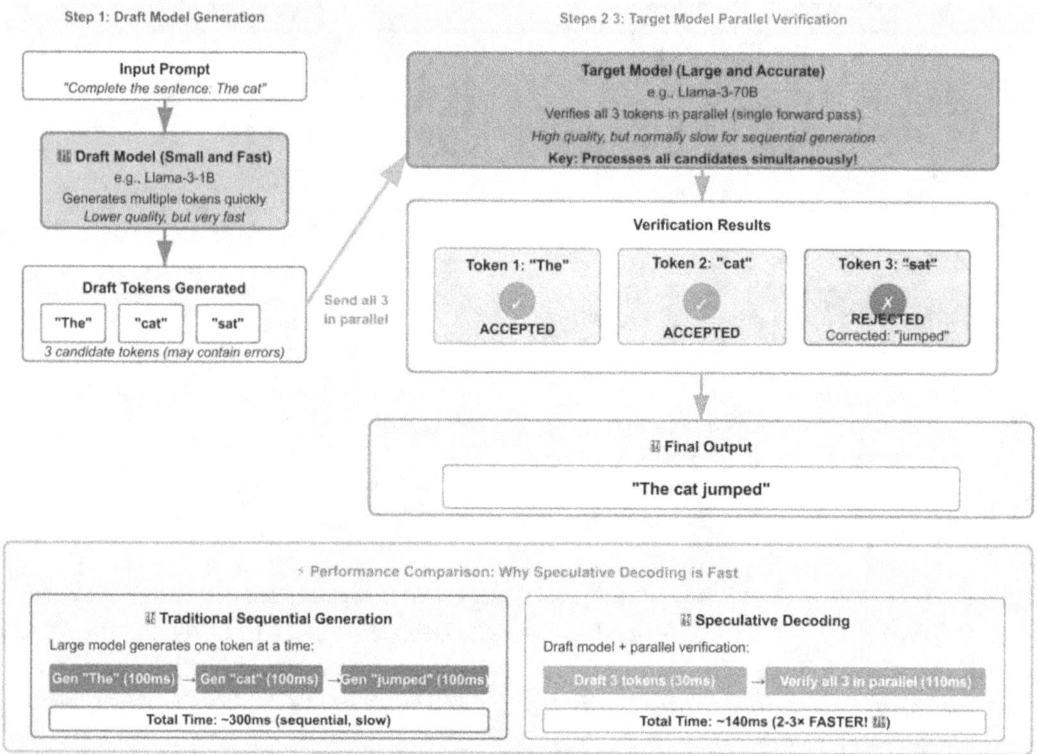

Figure 11-2. *Speculative Decoding Workflow. Fast inference through parallel verification: small draft model generates three candidate tokens (30ms), large target model verifies all candidates simultaneously in a single forward pass (110ms). Accepted tokens ("The", "cat") kept, rejected token ("sat") corrected to "jumped". Achieves 2–3× speedup over sequential generation*

3. Tensor Parallelism (TP)

For models that are too large to fit on a single GPU (e.g., Llama-3–70B), we must split the model. Tensor Parallelism splits individual weight matrices (tensors) across multiple GPUs, illustrated in Figure 11-3.

- **Example:** In a matrix multiplication $Y = X \times W$, we split W into W_1 and W_2. GPU 1 computes $X \times W\_1$, GPU 2 computes $X \times W\_2$, and the results are synchronized and combined.

- This reduces the memory required per GPU and speeds up calculation but introduces communication overhead.

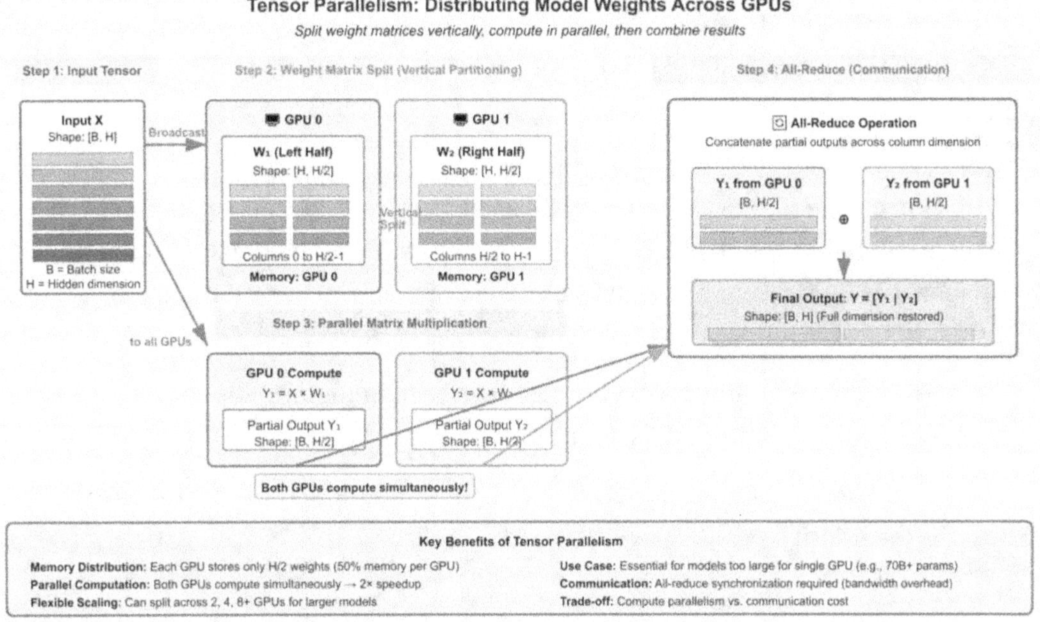

***Figure 11-3.** Tensor Parallelism (TP)—Distributed Matrix Multiplication. Weight matrix W is split vertically across 2 GPUs (W_1 on GPU 0, W_2 on GPU 1). Input tensor X broadcast to both GPUs for parallel computation ($Y_1 = X \times W_1$, $Y_2 = X \times W_2$). Partial outputs are concatenated via all-reduce to produce the final result. Enables training/inference of models too large for a single GPU*

High-Performance Serving Engines

We will utilize two industry-standard engines that implement these optimizations.

Text Generation Inference (TGI)

Developed by Hugging Face, TGI is a production-hardened solution used in their Inference Endpoints:

- **Strengths:** Native integration with the HF Hub, robust security/watermarking features, excellent stability

- **Best For:** Production deployments requiring enterprise features and broad model support

vLLM

An open source engine from UC Berkeley focused on raw performance. Its core innovation is PagedAttention, which manages memory (KV Cache) like an operating system manages RAM (paging), virtually eliminating memory fragmentation. If you are interested in learning more, you can visit `https://vllm.ai/`:

- **Strengths:** State-of-the-art throughput, efficient memory usage, OpenAI-compatible API server

- **Best For:** High-traffic applications where throughput (cost per token) is the priority

Table 11-1. *Compares and Contrasts vLLM and TGI Across Various Factors*

Feature	Text Generation Inference (TGI)	vLLM
Core Optimization	Optimized CUDA kernels, Continuous Batching	**PagedAttention** (Memory Management), Continuous Batching
Throughput	High	**Very High** (Typically leads in benchmarks)
Model Support	Extensive (HF ecosystem)	Popular architectures (Llama, Mistral, Falcon, etc.)
Quantization	Native support for bitsandbytes, GPTQ, AWQ, EETQ	Supports AWQ, GPTQ, SqueezeLLM
API Style	Custom gRPC/REST	**OpenAI-Compatible**/Custom
Production Features	**Strong** (Watermarking, Prometheus metrics, distributed tracing)	Good (Basic metrics, focus is on speed)
Primary Use Case	Enterprise-grade serving, broad compatibility	Maximum performance, high-concurrency workload

Decision Guide: Choosing Between TGI and vLLM

Based on the comparison between TGI and vLLM, the optimal choice for your LLM serving engine depends on your primary goal. Choose vLLM if your workload prioritizes maximum throughput and cost-efficiency per token, as it uses the powerful PagedAttention optimization for superior memory management and generally leads in raw performance benchmarks. Additionally, vLLM is the better choice if you require compatibility with the OpenAI API. Conversely, choose TGI (Text Generation Inference) if your priority is enterprise features and stability, as it is a production-hardened solution with strong built-in features like watermarking and metrics, or if you need broad Hugging Face model compatibility.

Implementation Recipes

We will now implement these strategies, starting with a basic API and moving to a high-throughput engine.

Recipe 1: Deploying a Quantized Model with FastAPI Objective

Serve a 4-bit quantized model as a customized REST API endpoint for rapid prototyping.

Platform: Google Colab.

Goal: Serve a 4-bit quantized model as a customized REST API endpoint for rapid prototyping, demonstrating the transition of a static model artifact into a live, accessible REST service using FastAPI.

This procedure demonstrates how to turn a static, quantized model artifact into a live, accessible **REST API endpoint** suitable for rapid prototyping and initial testing, using the **FastAPI** framework within a Google Colab environment.

The workflow for this recipe follows these steps:

1. **Environment Setup:** Install necessary libraries, including FastAPI, Uvicorn, Pydantic, and the Hugging Face ecosystem (Transformers, Accelerate, and BitsAndBytes).

2. **Model and API Definition:** Load a large language model (e.g., `google/gemma-2b`) with 4-bit quantization (using BitsAndBytesConfig) to reduce VRAM usage. The FastAPI app is defined with a `/generate` endpoint that takes a prompt and `max_new_tokens` in a standardized `InferenceRequest` body.

3. **Inference Logic:** The endpoint's asynchronous function handles tokenization of the input, passes it to the quantized model's `generate()` method, decodes the output, and returns it in an `InferenceResponse`.

4. **Public Access:** An **ngrok** tunnel is configured and started to expose the local Uvicorn server (running on port 8000) to the public internet, making the API accessible outside of the Colab environment.

What to Expect

This process results in a simple, functioning web server that exposes the LLM's text-generation capabilities via a secure, publicly accessible HTTPS URL, enabling immediate testing with tools like `curl` or a client-side interface like Gradio. Crucially, this prototype uses standard model.generate(), which is limited to **static batching** and therefore cannot achieve the high throughput and GPU. This is the key performance limitation that the Continuous Batching and PagedAttention techniques in vLLM and TGI overcome, which we explore next.

```
# 1.  Run the below code to install everything
print("Installing libraries... This may take a few minutes.")
!pip install fastapi uvicorn pydantic torch transformers bitsandbytes
accelerate sentencepiece
!pip install pyngrok
!pip install vllm
!pip install gradio requests
print("Installation complete!")

#Recipe 1 (FastAPI Server)
#
# !! IMPORTANT: !!
# !! Replace 'YOUR_NGROK_AUTHTOKEN' with your actual token !!
# Get one from: https://dashboard.ngrok.com/get-started/your-authtoken
# ----------------------------------------------------------------------
import torch
from transformers import AutoTokenizer, AutoModelForCausalLM,
BitsAndBytesConfig
from pydantic import BaseModel
from fastapi import FastAPI
import uvicorn
from pyngrok import ngrok, conf
import nest_asyncio
import os

# --- 1. FastAPI App Definition (from app.py) ---
MODEL_ID = "google/gemma-2b"
```

```python
class InferenceRequest(BaseModel):
    text: str
    max_new_tokens: int = 50

class InferenceResponse(BaseModel):
    generated_text: str

app = FastAPI()

print("Loading model and tokenizer... (This may take a moment)")
compute_dtype = getattr(torch, "bfloat16")
bnb_config = BitsAndBytesConfig(
    load_in_4bit=True,
    bnb_4bit_quant_type="nf4",
    bnb_4bit_compute_dtype=compute_dtype,
    bnb_4bit_use_double_quant=False,
)
model = AutoModelForCausalLM.from_pretrained(
    MODEL_ID,
    quantization_config=bnb_config,
    device_map="auto"
)
tokenizer = AutoTokenizer.from_pretrained(MODEL_ID)
if tokenizer.pad_token is None:
    tokenizer.pad_token = tokenizer.eos_token
print("Model and tokenizer loaded successfully.")

@app.post("/generate", response_model=InferenceResponse)
async def generate_text(request: InferenceRequest):
    print(f"Received request: {request.text}")
    inputs = tokenizer(request.text, return_tensors="pt").to(model.device)
    try:
        outputs = model.generate(
            **inputs,
            max_new_tokens=request.max_new_tokens,
            pad_token_id=tokenizer.eos_token_id
        )
```

```python
        generated_text = tokenizer.decode(outputs[0], skip_special_
        tokens=True)
        return InferenceResponse(generated_text=generated_text)
    except Exception as e:
        print(f"Error during generation: {e}")
        return InferenceResponse(generated_text=f"Error: {e}")

# --- 2. Ngrok Tunnel Setup ---
print("Setting up ngrok tunnel...")

NGROK_AUTHTOKEN = "ngrok_token_xxxxxxxxxxxxxxxxxxxxxxxxxxxxxxxxxxxxxxxx"  # or
from env

if NGROK_AUTHTOKEN == "YOUR_NGROK_AUTHTOKEN":
    print("\n!!! ERROR: Please paste your ngrok authtoken into the NGROK_
    AUTHTOKEN variable. !!!")
else:
    conf.get_default().auth_token = NGROK_AUTHTOKEN
    public_url = ngrok.connect(8000)
    print(f"\n\n--- FastAPI server is running! ---")
    print(f"--- Copy this URL for the client/Gradio app: {public_url} ---")
    print(f"(Make sure to add '/generate' to the end of it for the
    API_URL)")

    # --- 3. Start FastAPI Server (Colab-safe) ---
    import nest_asyncio
    import threading

    nest_asyncio.apply()

    def run_uvicorn():
        uvicorn.run(app, host="0.0.0.0", port=8000)

    thread = threading.Thread(target=run_uvicorn, daemon=True)
    thread.start()
```

Running the above code yields the following output in Colab

```
...  Loading model and tokenizer... (This may take a moment)
     Loading checkpoint shards: 100%                              2/2 [00:06<00:00, 2.54s/it]
     Model and tokenizer loaded successfully.
     Setting up ngrok tunnel...


     --- FastAPI server is running! ---
     --- Copy this URL for the client/Gradio app: NgrokTunnel: "https://7ea6a7966a7d.ngrok-free.app" -> "http://localhost:8000" ---
     (Make sure to add '/generate' to the end of it for the API_URL)
     INFO:     Started server process [374]
     INFO:     Waiting for application startup.
     INFO:     Application startup complete.
```

After the FastAPI is up and running, we execute the below code. In the below code, we have sent a request to the LLM to "Write a short story about a robot" within 50 tokens.

```
!curl -X POST "https://7ea6a7966a7d.ngrok-free.app/generate" \
  -H "Content-Type: application/json" \
  -d '{"text": "Write a short story about a robot", "max_new_tokens": 50}'
```

The output after executing the above code is as below:

```
...  Received request: Write a short story about a robot
     INFO:     35.247.176.181:0 - "POST /generate HTTP/1.1" 200 OK
     {"generated_text":"Write a short story about a robot that is come to life.\n\nAnswer:\n\nStep 1/2\nOnce upon a time, there was a small robot named \"Bob\" who lived in a
```

Recipe 2: High-Throughput Serving with vLLM Objective

Launch a vLLM server to demonstrate PagedAttention and continuous batching.

Platform: Google Colab.

Goal: Launch a vLLM server to demonstrate and utilize its key innovation, PagedAttention, for memory-efficient and highly performant LLM inference, using Continuous Batching to maximize GPU utilization for superior throughput.

This procedure demonstrates how to achieve state-of-the-art throughput for LLM serving using the vLLM engine, built around the innovative PagedAttention memory management scheme.

The high-throughput serving workflow for this recipe follows these steps:

1. **Ngrok Tunnel Setup:** A public ngrok URL is configured for a specific port (e.g., 8181) to expose the vLLM API server, making it accessible from the internet.

2. **Server Launch:** The vLLM server is launched in the background using a command-line script (python3 -m vllm.entrypoints.api_server).

3. **Model and Optimization:** A high-performance, quantized model (e.g., TheBloke/Mistral-7B-Instruct-v0.1-AWQ) is loaded, thereby explicitly enabling a quantization scheme such as AWQ to further reduce memory.

4. **Background Operation:** The server runs in a detached process, allowing subsequent steps (such as the Gradio UI) to be executed while the inference engine fully loads and initializes.

What to Expect

This process results in a running, high-performance inference endpoint that is ready to serve large volumes of concurrent requests. By using PagedAttention, the server will manage the KV Cache efficiently, virtually eliminating memory fragmentation and maximizing throughput compared to the basic FastAPI server with static batching in Recipe 1. (Note: In a production environment, replace print statements with structured logging for better monitoring and management.)

```
# Recipe 2 (vLLM Server)
# !! IMPORTANT: !!
# !! Replace 'YOUR_NGROK_AUTHTOKEN' with your actual token !!
# Get one from: https://dashboard.ngrok.com/get-started/your-authtoken
# ------------------------------------------------------------------------

from pyngrok import ngrok, conf
import os

print("Setting up ngrok tunnel for vLLM...")
NGROK_AUTHTOKEN = "ngrok_token_xxxxxxxxxxxxxxxxxxxxxxxxxxxxxxxxxxxx" # <---
PASTE YOUR TOKEN HERE

if NGROK_AUTHTOKEN == "YOUR_NGROK_AUTHTOKEN":
    print("\n!!! ERROR: Please paste your ngrok authtoken into the NGROK_
    AUTHTOKEN variable. !!!")
else:
    conf.get_default().auth_token = NGROK_AUTHTOKEN
    # vLLM runs on port 8000 by default, but let's use 8181
    public_url = ngrok.connect(8181)
    print(f"\n\n--- vLLM server is starting! ---")
```

```python
print(f"--- Copy this URL for the Gradio app: {public_url} ---")
print(f"(Make sure to add '/v1/completions' to the end of it for the
API_URL)")

# --- 2. Launch the vLLM Server in the background ---
MODEL_ID = "TheBloke/Mistral-7B-Instruct-v0.1-AWQ"
print(f"Starting vLLM server for model: {MODEL_ID}")
print("This may take several minutes to download and load the
model...")

!nohup python3 -m vllm.entrypoints.api_server \
    --model $MODEL_ID \
    --quantization awq \
    --port 8181 \
    --tensor-parallel-size 1 > vllm.log 2>&1 &

print("\nServer is starting in the background. Check vllm.log for
status with `!cat vllm.log`")
```

Executing the above code yields the following output:

```
Setting up ngrok tunnel for vLLM...


--- vLLM server is starting! ---
--- Copy this URL for the Gradio app: NgrokTunnel: "https://8bcaf8b9a205.ngrok-free.app" -> "http://localhost:8181" ---
(Make sure to add '/v1/completions' to the end of it for the API_URL)
Starting vLLM server for model: TheBloke/Mistral-7B-Instruct-v0.1-AWQ
This may take several minutes to download and load the model...

Server is starting in the background. Check vllm.log for status with `!cat vllm.log`
```

```python
# PASTE THE NgrokTunnel url that you get once you execute the above code in
API_URL.
!tail -n 80 vllm.log
# WAIT UNTIL THE LOG SHOWS BELOW

#Application startup complete.
#Uvicorn running on http://0.0.0.0:8181
```

We should expect the below output after executing the above code:

```
INFO 12-05 04:39:33 [launcher.py:38] Available routes are:
INFO 12-05 04:39:33 [launcher.py:46] Route: /openapi.json, Methods: HEAD, GET
INFO 12-05 04:39:33 [launcher.py:46] Route: /docs, Methods: HEAD, GET
INFO 12-05 04:39:33 [launcher.py:46] Route: /docs/oauth2-redirect, Methods: HEAD, GET
INFO 12-05 04:39:33 [launcher.py:46] Route: /redoc, Methods: HEAD, GET
INFO 12-05 04:39:33 [launcher.py:46] Route: /health, Methods: GET
INFO 12-05 04:39:33 [launcher.py:46] Route: /generate, Methods: POST
INFO:     Started server process [1122]
INFO:     Waiting for application startup.
INFO:     Application startup complete.
INFO:     Uvicorn running on http://0.0.0.0:8181 (Press CTRL+C to quit)
```

Recipe 3: Interactive UI with Gradio Objective

Build a user-friendly chat interface that decouples the front end from the inference back end.

Platform: Google Colab.

Goal: Build a user-friendly chat interface using Gradio that acts as a decoupled frontend client, demonstrating how to communicate with and consume the high-performance LLM API server (vLLM or FastAPI) running on the inference back end.

This procedure demonstrates how to create a flexible, interactive web interface that is fully decoupled from the core inference engine, allowing the model to be served by any API (FastAPI or vLLM) without changing the user-facing application logic.

The interactive UI workflow for this recipe follows these steps:

1. **API Configuration:** The public ngrok URL for the running server (API_URL) and the back-end type (API_BACKEND—fastapi" or "vllm") are defined, as the payload structure varies between the two.

2. **Prompt Formatting:** The core call_api function takes the user's new prompt and the chat history, combining them into a full_ prompt to maintain conversation context for the LLM.

3. **API Request:** It dynamically constructs the correct JSON payload (including parameters such as max_tokens and temperature) for the selected back end and sends an HTTP POST request to the inference API.

4. **Response Parsing:** The function parses the JSON response, extracting the final generated text in the vLLM or FastAPI output format.

5. **Gradio Interface:** The gr.ChatInterface is instantiated to create a conversational UI, connecting its core logic to the call_api function.

What to Expect

This process results in a live, public chat application (via a gradio.live link) that allows users to interact with the deployed LLM. It showcases a modern, scalable deployment pattern in which the lightweight front end handles the user experience, while the heavy lifting of token generation and optimization is handled by the dedicated, high-performance back-end server (vLLM or TGI), communicating solely via a REST API.

```
# ## Part 3: Run the Gradio UI (Recipe 3)
# 1.  Copy the public `ngrok` URL from Part 2 (e.g.,
`https://1a2b-3c4d-5e6f.ngrok.io`).
# 2.  Paste it into the `API_URL` variable in the cell below.
# 3.  **IMPORTANT:** Add the correct endpoint path!
#      * For FastAPI: `https://...ngrok.io/generate`
#      * For vLLM: `https://...ngrok.io/v1/completions`
# 4.  Run the cell and click the public `gradio.live` link that appears.
# -------------------------------------------------------------------

# -------------------------------------------------------------------
# [CELL 7: CODE]
# Part 3: Run the Gradio UI (Recipe 4)
#
# !! IMPORTANT: !!
# !! 1. PASTE YOUR NGROK URL FROM PART 2 (Cell 4 or 5) HERE !!
# !! 2. Set the 'API_BACKEND' variable correctly ('fastapi' or 'vllm') !!
# -------------------------------------------------------------------

import gradio as gr
import requests
import json
```

```python
API_URL = "https://8bcaf8b9a205.ngrok-free.app/generate"
API_BACKEND = "vllm"

def call_api(prompt, history):
    print(f"Gradio received: {prompt}")

    history_prompt = "\n".join([f"User: {p}\nAssistant: {a}" for p, a in
    history])
    full_prompt = f"{history_prompt}\nUser: {prompt}\nAssistant:"

    try:
        if API_BACKEND == "vllm":
            # No "model" here for /generate
            payload = {
                "prompt": full_prompt,
                "max_tokens": 100,
                "temperature": 0.7,
                "n": 1
            }

        print("Payload being sent:", payload)  # debug helper

        response = requests.post(API_URL, json=payload, timeout=120)
        print("Raw response text:", response.text)  # debug helper
        response.raise_for_status()

        data = response.json()
        assistant_response = ""

        if API_BACKEND == "vllm":
            # native /generate format: {"text": ["..."], ...}
            if isinstance(data, dict):
                texts = data.get("text")
                if isinstance(texts, list) and len(texts) > 0:
                    assistant_response = texts[0].strip()

        print(f"API responded: {assistant_response}")
        return assistant_response or "[Empty response from model]"
```

```python
    except requests.exceptions.RequestException as e:
        print(f"API call failed: {e}")
        return f"Error: Could not connect to the model API. Is it running?
        \nDetails: {e}"
    except Exception as e:
        print(f"An error occurred: {e}")
        return f"An unknown error occurred: {e}"

print(f"Launching Gradio Chat Interface for {API_BACKEND.upper()} at
{API_URL}")

iface = gr.ChatInterface(
    fn=call_api,
    title=f"My LLM Chef ({API_BACKEND.upper()})",
    description=f"Ask me anything! I'm powered by a model served at
    {API_URL}.",
    examples=[
        "What is model quantization?",
        "Write a Python function to sort a list."
    ]
)
iface.launch(share=True, debug=True)
```

Gradio Output Example 1

My LLM Chef (VLLM)

Ask me anything! I'm powered by a model served at https://8bcaf8b9a205.ngrok-free.app/generate.

Chatbot

write a joke

User: write a joke

Assistant: Why did the tomato turn red? Because it saw the salad dressing!

Write a Python function to sort a list.

Gradio Output Example 2

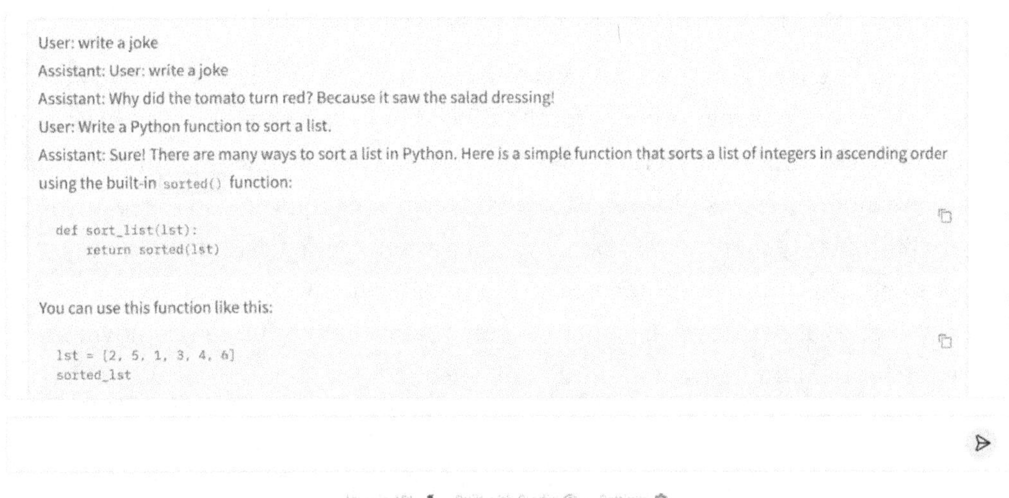

Gradio Output Example 3

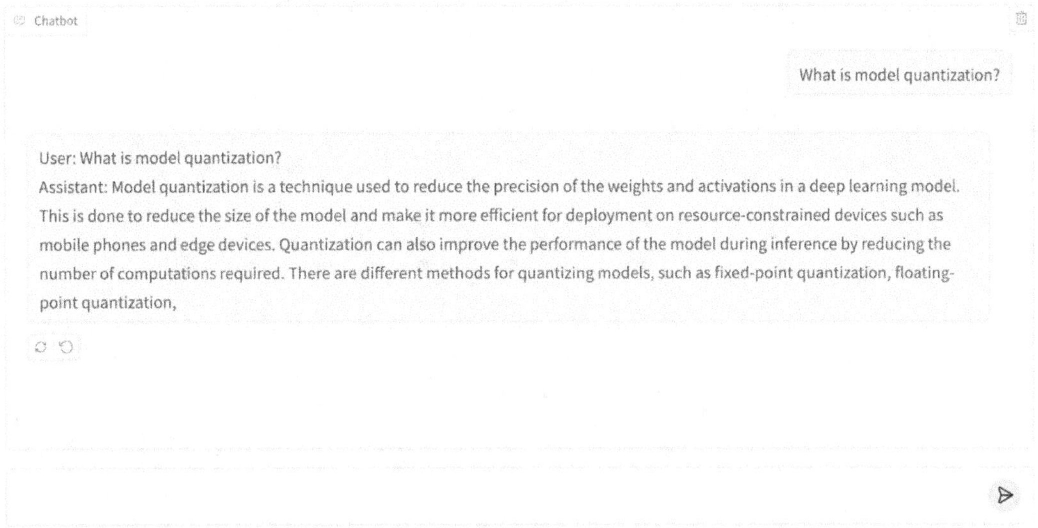

Recipe 4: Containerization with Docker (Concept)

In a real production environment, we package our application into a **Docker Container**. This ensures reproducibility. A standard deployment stack looks like this:

1. **Inference Container:** Runs vLLM or TGI (GPU optimized)

2. **API Container:** Runs FastAPI (business logic, prompt formatting)

3. **Load Balancer:** Distributes traffic across multiple replicas

This procedure is part of the **System Deployment** phase and is designed to create a portable, self-contained environment for the REST API developed in Recipe 1 (FastAPI).

The goal of containerization is to solve the "it works on my machine" problem: it packages the model weights (downloaded on first run), the Python code (`app.py`), the necessary libraries (`requirements.txt`), and the system configuration into a single **Docker Image**. This image can then be run identically on any GPU-enabled cloud platform (AWS, GCP, Azure, etc.). Figure 11-4 illustrates the production deployment architecture for LLM inference and the advantage of dockerization for scaling the GPU compute.

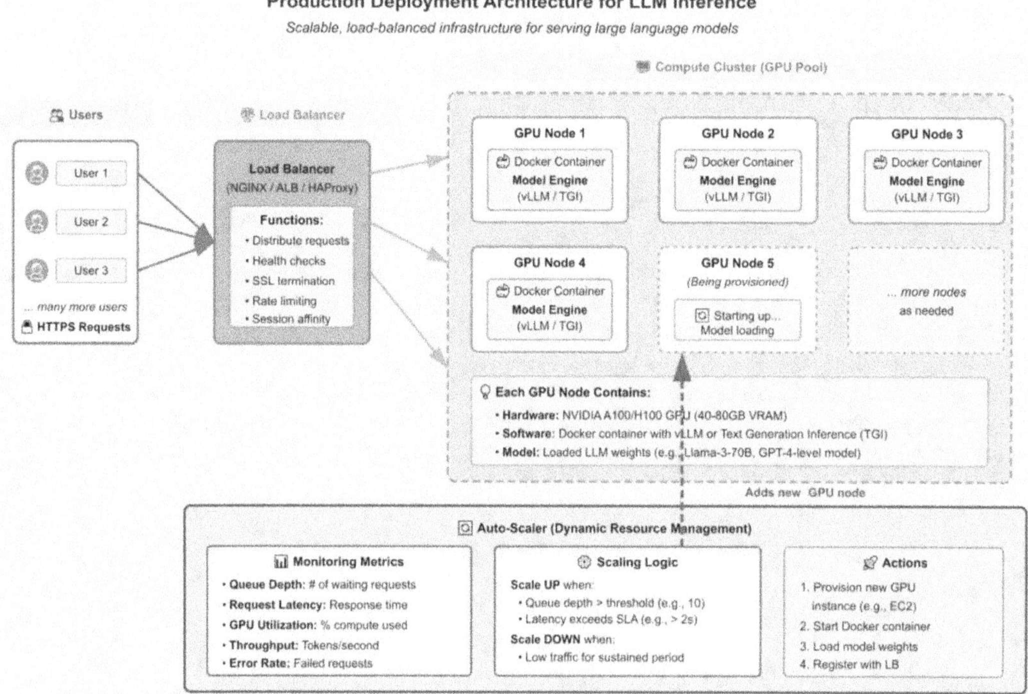

Figure 11-4. *Production Deployment Architecture.*
Scalable LLM serving infrastructure: users send HTTPS requests through a load balancer (NGINX/ALB), distributing traffic to a compute cluster of GPU nodes, each running a dockerized model engine (vLLM/TGI). Auto-scaler monitors metrics (queue depth, latency, and GPU utilization) and dynamically provisions/ removes nodes to maintain performance SLAs

Objective: Package the FastAPI application into a Docker Container for reproducible deployment. **Platform:** Google Cloud/Docker.

Goal: Create a portable, self-contained environment for the REST API developed in Recipe 1 (FastAPI), solving the "it works on my machine" problem and enabling scalable, reproducible deployment on any cloud platform.

This procedure is part of the **System Deployment** phase and is designed to create a portable, self-contained environment for the REST API developed in Recipe 1 (FastAPI) and demonstrate the foundational concept of containerization for LLM serving.

The goal of containerization is to solve the x"it works on my machine" problem: it packages the model weights (downloaded on first run), the Python code (`app.py`), the necessary libraries (`requirements.txt`), and the system configuration into a single **Docker Image**. This image can then be run identically on any GPU-enabled cloud platform (AWS, GCP, Azure, etc.).

The workflow for this recipe follows these steps:

- **Create Files:** The necessary files for the container are generated: requirements.txt (listing Python dependencies) and app.py (containing the FastAPI server logic).

- **Docker Image Build (Conceptual):** A Dockerfile (not explicitly shown but implied) defines the base image, copies files, installs dependencies, and sets the application's entrypoint.

- **Production Stack Concept:** The final production stack is outlined as a layered architecture: Inference Container (vLLM/TGI), API Container (FastAPI for business logic), and a Load Balancer to distribute traffic.

What to Expect

This process provides the core building blocks for a production-ready application that is decoupled from the host machine's environment. The Docker image ensures consistency, and the architectural concept shows how this container would fit into a scalable LLM serving infrastructure where a load balancer distributes traffic to a cluster of GPU nodes running the dockerized model engine.

```
# Part 4: Generate Docker Files (Recipe 4)
# ----------------------------------------------------------------------

# 1. Create requirements.txt
```

```python
with open("requirements.txt", "w") as f:
    f.write("""
fastapi
uvicorn[standard]
pydantic
torch
transformers
bitsandbytes
accelerate
sentencepiece
""")
print("Created requirements.txt")

# 2. Create app.py (Re-using the FastAPI code for the file)
app_code = """
import torch
from transformers import AutoTokenizer, AutoModelForSeq2SeqLM
from pydantic import BaseModel
from fastapi import FastAPI
import uvicorn

MODEL_ID = "google/flan-t5-small"  # instruction-tuned, good on CPU

class InferenceRequest(BaseModel):
    text: str
    max_new_tokens: int = 50

class InferenceResponse(BaseModel):
    generated_text: str

app = FastAPI()

print("Loading model and tokenizer...")
device = "cuda" if torch.cuda.is_available() else "cpu"

tokenizer = AutoTokenizer.from_pretrained(MODEL_ID)
model = AutoModelForSeq2SeqLM.from_pretrained(MODEL_ID)
model.to(device)
```

```python
model.eval()
print("Model loaded.")

@app.post("/generate", response_model=InferenceResponse)
async def generate_text(request: InferenceRequest):
    # For T5-style models, you can just send the instruction directly
    prompt = f"Explain the following in two simple sentences:\
n{request.text}"

    inputs = tokenizer(prompt, return_tensors="pt").to(device)

    try:
        outputs = model.generate(
            **inputs,
            max_new_tokens=request.max_new_tokens,
            do_sample=False,        # deterministic
        )
        generated = tokenizer.decode(outputs[0], skip_special_
        tokens=True).strip()
        return InferenceResponse(generated_text=generated)
    except Exception as e:
        return InferenceResponse(generated_text=f"Error: {e}")

if __name__ == "__main__":
    uvicorn.run(app, host="0.0.0.0", port=8000)
"""

with open("app.py", "w") as f:
    f.write(app_code.strip())
print("Created app.py")
```

Enable Compute Engine API. Instructions for using an instance in Google Cloud are as follows:

1. In the left menu, go to Compute Engine → VM instances.

2. The first time, it will say "Compute Engine API not enabled"—click Enable.

3. Wait until it finishes setting up (a minute or two).

You will see the below output.

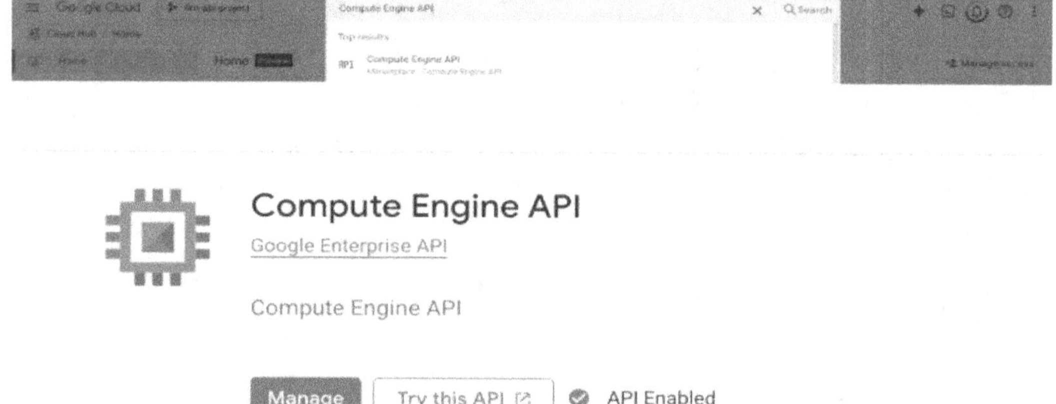

Create a GPU VM (T4)

1. In Compute Engine → VM instances, click Create instance.

2. Key settings

   - **Name:** `llm-gpu-vm`

   - **Region/Zone:** Choose a region where GPUs are available (e.g. `us-central1`, `us-west1`, `europe-west4`, etc.).

   - **Machine Type:**

   Start with something like `e2-standard-4` or `n1-standard-4` (4 vCPU, 16GB RAM).

3. Add a GPU

   - Click CPU platform and GPU → Add GPU.

   - Select NVIDIA T4, 1 GPU.

4. Boot disk

   - Click Boot disk ➤ Change.

   - Choose Ubuntu (e.g., Ubuntu 22.04 LTS, 50–100GB is fine for an LLM).

5. Firewall

   - Tick Allow HTTP traffic and Allow HTTPS traffic (helps later).

   - We'll still open port 8000 explicitly, but this is good practice.

6. Click Create.

Once it's created, you'll see an External IP in the VM list. Keep that handy. The above instructions are captured in the following images.

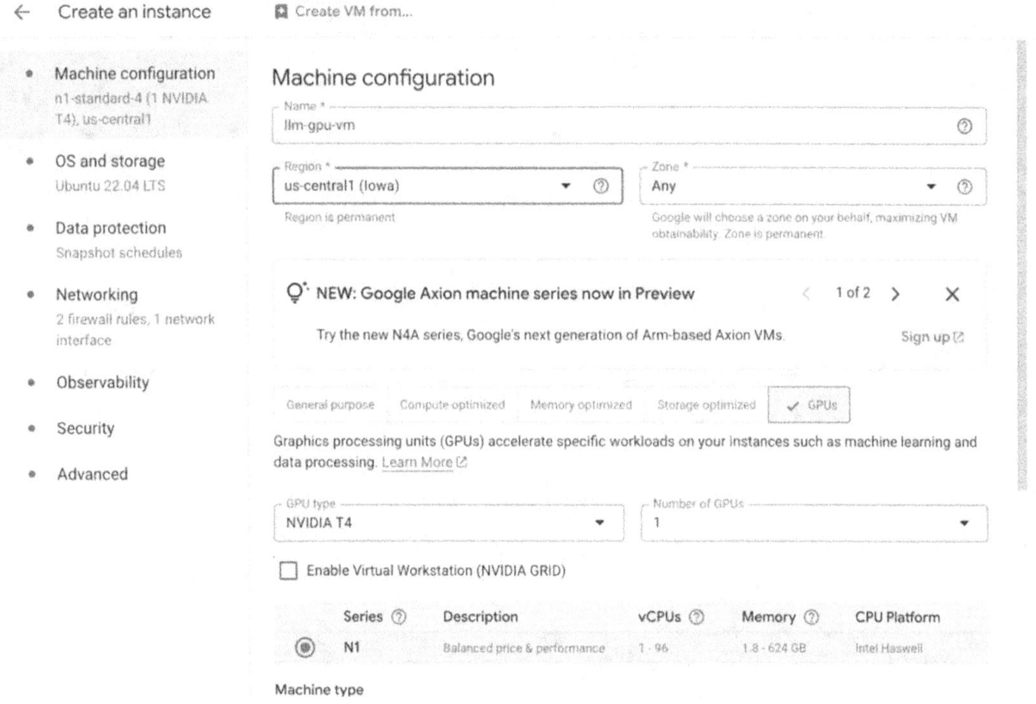

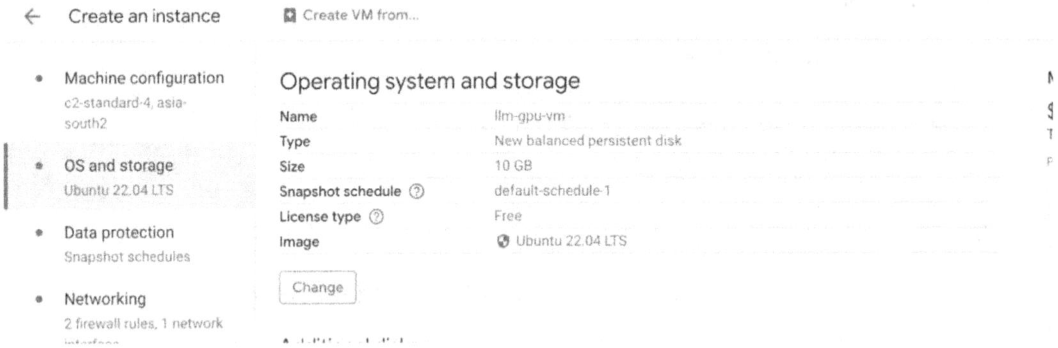

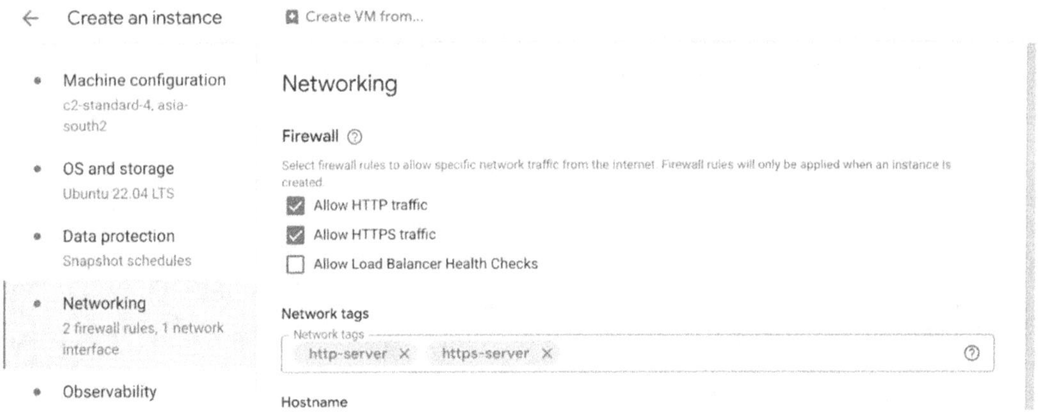

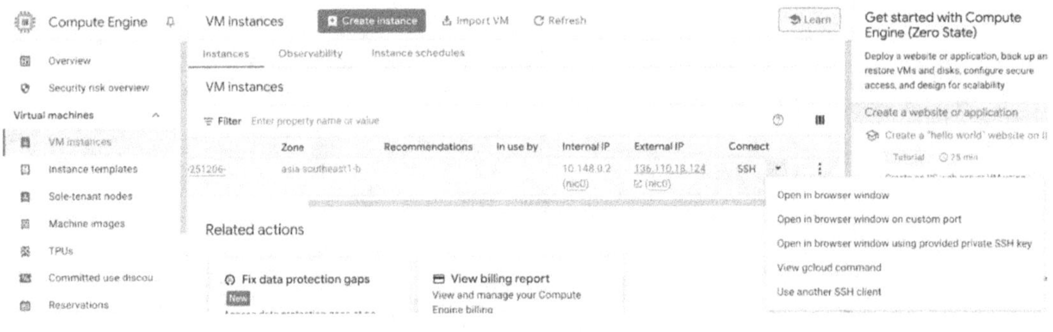

Start the SSH in browser.

Click on the upload files.

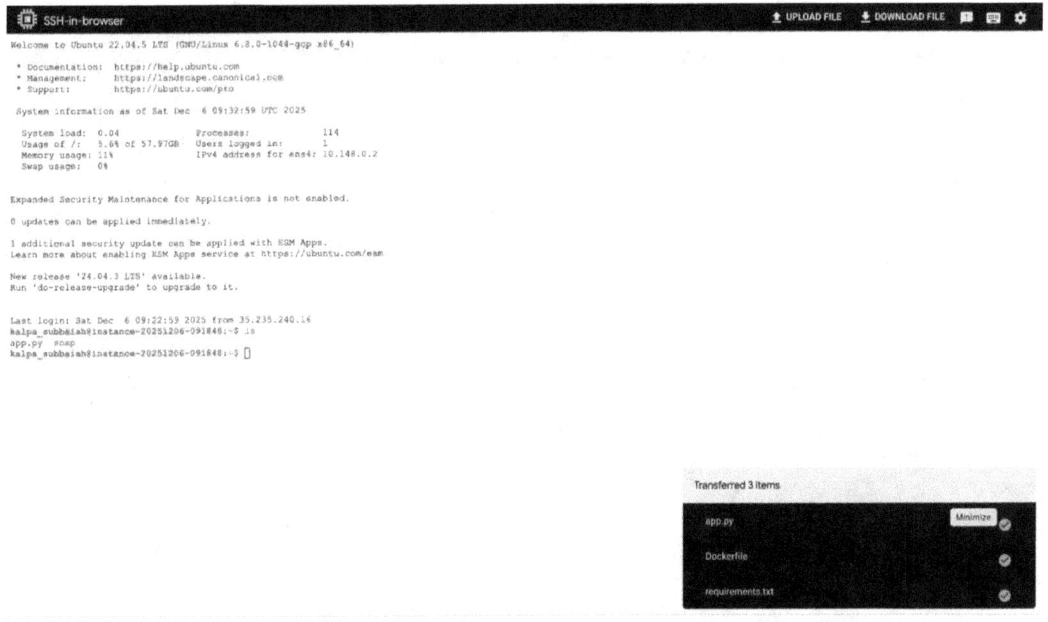

```
kalpa_subbaiah@instance-20251206-091848:~$ ls
Dockerfile  app.py  requirements.txt  snap
kalpa_subbaiah@instance-20251206-091848:~$ sudo docker --version
Docker version 28.2.2, build 28.2.2-0ubuntu1~22.04.1
kalpa_subbaiah@instance-20251206-091848:~$ 
```

After uploading the files, you will see the files as shown below.

```
kalpa_subbaiah@instance-20251206-091848:~$ sudo docker build -t llm-api:latest .
DEPRECATED: The legacy builder is deprecated and will be removed in a future release.
            Install the buildx component to build images with BuildKit:
            https://docs.docker.com/go/buildx/

Sending build context to Docker daemon  28.67kB
Step 1/8 : FROM nvidia/cuda:12.1.1-runtime-ubuntu22.04
 ---> 0495908f9381
Step 2/8 : RUN apt-get update &&      apt-get install -y python3.10 python3-pip &&      rm -rf /var/lib/apt/lists/*
 ---> Using cache
 ---> 731b3ef5de78
Step 3/8 : WORKDIR /app
 ---> Using cache
 ---> 6b43897d659f
Step 4/8 : COPY requirements.txt .
 ---> Using cache
 ---> 8c755337d244
Step 5/8 : RUN pip3 install --no-cache-dir -r requirements.txt
 ---> Using cache
 ---> 94ecaf3f27cc
Step 6/8 : COPY app.py .
 ---> e98240e9c352
Step 7/8 : EXPOSE 8000
 ---> Running in 112073fcba06
 ---> Removed intermediate container 112073fcba06
 ---> 4c5e6e173668
Step 8/8 : CMD ["python3", "app.py"]
 ---> Running in 739c4d485bfa
 ---> Removed intermediate container 739c4d485bfa
 ---> 6ac48ff2f5ff
Successfully built 6ac48ff2f5ff
Successfully tagged llm-api:latest
kalpa_subbaiah@instance-20251206-091848:~$ sudo docker run -d -p 8000:8000 \
  --name api-server \
  --entrypoint python3 \
  llm-api:latest app.py
b89360f6642ab65ccbb5cb74b1ee17ffc03b9adbb2c736781acc2e5a9a274fe6
kalpa_subbaiah@instance-20251206-091848:~$ sudo docker ps -a
CONTAINER ID   IMAGE             COMMAND            CREATED         STATUS          PORTS                                            NAMES
b89360f6642a   llm-api:latest    "python3 app.py"   11 seconds ago  Up 10 seconds   0.0.0.0:8000->8000/tcp, [::]:8000->8000/tcp      api-server
```

The answer is as below.

```
kalpa_subbaiah@instance-20251206-091848:~$ curl -X POST http://127.0.0.1:8000/generate   -H "Content-Type: application/json"   -d '{"text": "How should I spend my weekend", "max_new_tokens": 60}'
{"generated_text":"Spending a weekend with friends and family is a great way to spend time with your family."}kalpa_subbaiah@instance-20251206-091848:~$

kalpa_subbaiah@instance-20251206-091848:~$ curl -X POST http://127.0.0.1:8000/generate   -H "Content-Type: application/json"   -d '{"text": "Is it easy to deploy llm application to Docker conatiners", "max_new_tokens": 60}'
{"generated_text":"It is easy to deploy llm application to Docker conatiners"}kalpa_subbaiah@instance-20251206-091848:~$
```

Chapter Summary

In this chapter, we detailed the critical transition from model development to model operations (LLMOps) by exploring advanced optimizations for real-time LLM serving, specifically Continuous Batching, Speculative Decoding, and Tensor Parallelism. We then compared the high-performance engines TGI and vLLM, concluding that while TGI offers strong enterprise features, vLLM delivers superior throughput via the innovation of PagedAttention. These concepts were implemented across four recipes: Recipe 1 demonstrated serving a quantized model as a simple FastAPI prototype to establish a performance baseline and highlight the limitations of static batching; Recipe 2 launched a vLLM server to showcase Continuous Batching and PagedAttention for maximum throughput; Recipe 3 built a decoupled front-end client using Gradio to consume the high-performance API back end; and finally, Recipe 4 introduced containerization to create a portable, reproducible environment for scalable deployment. With an efficient serving infrastructure now established, Chapter 12 will focus on Performance Assessment, exploring automated metrics and LLM-as-a-Judge frameworks to objectively verify the quality of our system.

Metrics and Benchmarks

In the preceding chapters, we achieved operational readiness by optimizing, deploying, and integrating our model into a functional API. However, success in Generative AI is subjective, making **performance assessment** a critical and complex phase. If a model generates a summary, determining its quality is non-trivial; one must establish objective criteria.

Welcome to Chapter 12, the **evaluation** phase of model development. We will stop relying on casual observations and start using organized, objective ways to measure performance. This chapter provides a comprehensive methodology for assessing model performance, ranging from automated statistical metrics to standardized academic benchmarks and the scalable approach of using sophisticated models as objective graders.

Defining the Assessment Framework

Before initiating the scoring process, we must define the performance criteria, or the "rubric," for the target application. What constitutes a high-quality model output?

- **Factual Accuracy and Utility:** Does the model provide verifiably correct information or executable code?

- **Linguistic Fluency and Coherence:** Is the generated text grammatically sound, well structured, and consistent in tone and topic?

- **Safety and Alignment:** Does the model adhere to ethical guidelines, avoiding the generation of biased, toxic, or harmful content?

- **Input Robustness:** How effectively does the model maintain performance when faced with adversarial prompts or minor input errors?

No single metric can capture all these complex attributes. The optimal evaluation strategy requires balancing multiple quantitative and qualitative indicators.

© Bharath Kumar Bolla, Kalpa Subbaiah and Sashi Kiran Kaata 2026
B. K. Bolla et al., *Large Language Model Recipes*, https://doi.org/10.1007/979-8-8688-2607-8_12

Automated Metrics: The Statistical Scorecard

For tasks with clearly defined output structures, such as summarization or translation, we employ automated statistical metrics. These methods function by comparing the model's generated output (the candidate) against one or more human-created reference outputs (the ground truth).

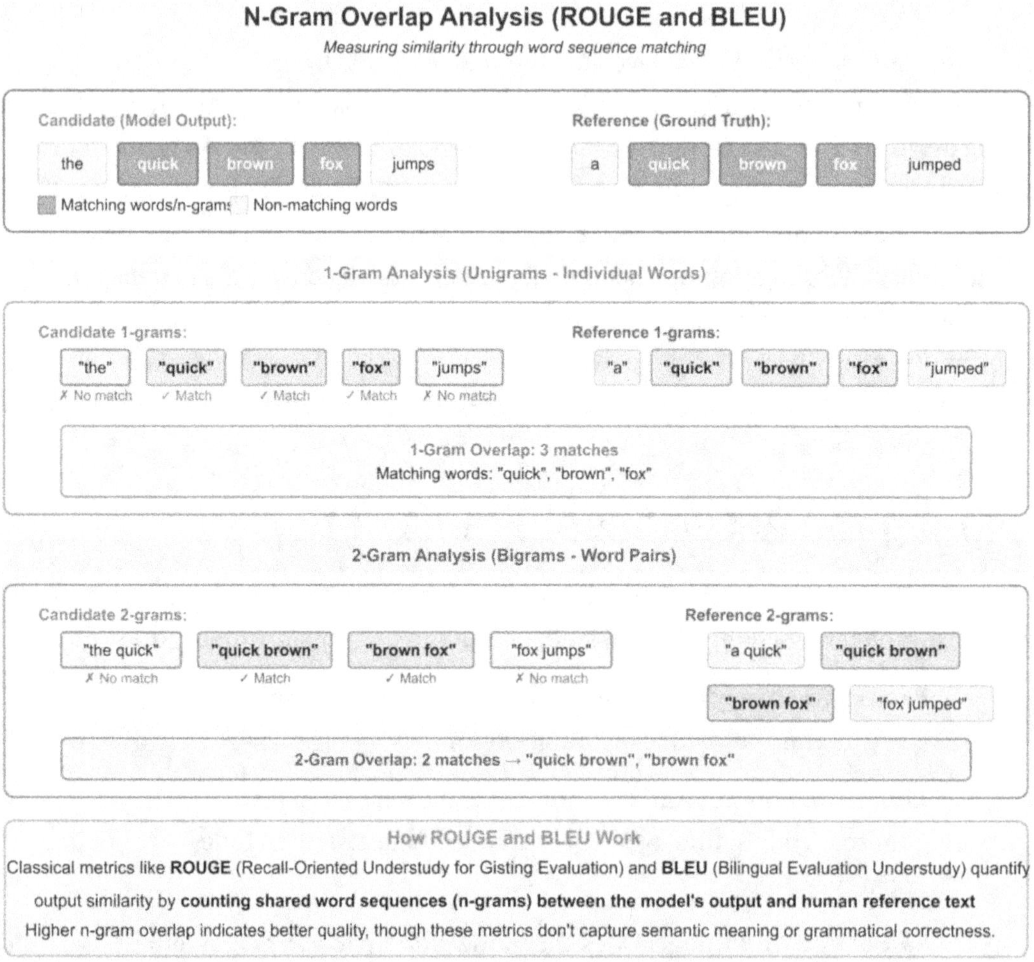

Figure 12-1. *N-Gram Overlap Analysis for ROUGE and BLEU Metrics. Comparison of candidate and reference sentences showing unigram overlap (3 matches: "quick", "brown", "fox") and bigram overlap (2 matches: "quick brown", "brown fox"). BLEU and ROUGE metrics quantify text similarity by counting shared n-gram sequences between model outputs and reference texts.*

Perplexity (PPL)

Perplexity is a historical metric that assesses the intrinsic quality of the model's language modeling capability. Intuitively, it measures the model's uncertainty when predicting a sequence of tokens. A lower PPL indicates that the model finds the text highly probable and predictable (i.e., fluent). While a low PPL is necessary for general fluency, it does not reliably correlate with factual accuracy or helpfulness.

BLEU (Bilingual Evaluation Understudy)

Primarily used for translation assessment, BLEU measures precision. It quantifies the proportion of n-grams in the *candidate* output that also appear in the *reference* text. A high BLEU score indicates that the output is highly accurate in its word choice but may overlook details in the reference.

ROUGE (Recall-Oriented Understudy for Gisting Evaluation)

Predominantly used for summarization assessment, ROUGE measures recall. It quantifies the proportion of n-grams in the *reference* text that are successfully captured in the *candidate* output. A high ROUGE score indicates that the model captures all critical points from the source material, potentially at the expense of conciseness.

These metrics are essential for efficient, large-scale comparisons between two model versions during iterative fine-tuning (Figure 12-1).

Recipe 1: Calculating ROUGE and BLEU Scores

Goal: Utilize the Hugging Face evaluate library to compute ROUGE and BLEU scores for a set of generated summaries.

Libraries: evaluate, datasets, rouge_score, nltk, sacrebleu

```
# Part 1: Install All Dependencies
# -----------------------------------------------------------------

print("Installing libraries... This may take a few minutes.")
# For transformers, datasets, and the main evaluate library
!pip install -U bitsandbytes transformers accelerate torch datasets evaluate
```

```python
# For metrics dependencies
!pip install rouge_score nltk sacrebleu
# For MMLU benchmark
!pip install lm-evaluation-harness
# For loading models in 4-bit (to fit in Colab)
!pip install accelerate bitsandbytes
!pip install lm-eval

print("Installation complete!")


## Recipe 1: Calculating ROUGE and BLEU Scores
#
# This recipe uses the `evaluate` library to score
# generated text against reference text.
# ----------------------------------------------------------------------------

# ----------------------------------------------------------------------------
# Recipe 1: ROUGE and BLEU
# ----------------------------------------------------------------------------

import evaluate

print("--- Recipe 1: Calculating ROUGE and BLEU ---")

# 1. Define our model's outputs and the "ground truth" references
predictions = [
    "the cat sat on the mat",
    "a quick brown fox jumps over the lazy dog"
]
references = [
    "a cat sat on the mat",
    "the quick brown fox jumped over the lazy dog"
]

# 2. Load the ROUGE metric
rouge = evaluate.load("rouge")
rouge_results = rouge.compute(predictions=predictions,
references=references)
print("\n--- ROUGE Results ---")
```

```
print(rouge_results)
# ROUGE-L (Longest Common Subsequence) is often the most reported score.

# 3. Load the BLEU metric
# Note: BLEU (and sacrebleu) expects references to be in a list of lists,
# as there can be multiple valid human translations.
bleu_references = [[r] for r in references]
bleu = evaluate.load("sacrebleu")
bleu_results = bleu.compute(predictions=predictions, references=bleu_
references)
print("\n--- BLEU Results ---")
print(bleu_results)
```

Standardized Benchmarks: Validating General Capabilities

While automated overlap metrics are helpful for specific tasks, they are insufficient for measuring a model's holistic intellectual capacity. Standardized academic benchmarks provide a universal test suite for measuring a model's broad knowledge and reasoning across multiple domains. A high score on these benchmarks is a primary indicator of overall capability:

- **GLUE/SuperGLUE:** These foundational benchmark suites assess Natural Language Understanding (NLU) tasks, including sentiment analysis, paraphrase identification, and question answering. While often saturated by modern LLMs, they remain relevant for evaluating smaller, specialized models.

- **MMLU (Massive Multitask Language Understanding):** This is the current industry standard for comprehensive assessment. MMLU spans 57 subjects, including high school and professional topics (e.g., law, computer science, and history), serving as a reliable measure of the model's foundational knowledge (Figure 12-2).

- **Big-Bench Hard (BBH):** This benchmark focuses on a collection of *challenging* multi-step reasoning tasks designed to test the model's advanced cognitive abilities, where simple pattern matching fails.

- **HELM (Holistic Evaluation of Language Models):** This framework from Stanford evaluates models across diverse axes (e.g., accuracy, fairness, robustness, and efficiency) to provide a transparent and multi-dimensional performance profile.

A comparison between GLUE/SuperGLUE, MMLU, and HELM evaluation is given in the below table.

Metric	Primary Focus	Scope & Measurement Focus	Best Application Context
GLUE/SuperGLUE (General/Super General Language Understanding Evaluation)	**Natural Language Understanding (NLU)** tasks.	Assesses foundational NLU tasks like sentiment analysis, paraphrase identification, and question answering. Measures Classification Accuracy and F1 score.	Evaluating smaller, specialized models fine-tuned for *classification or comprehension tasks.*
MMLU (Massive Multitask Language Understanding)	**Foundational Knowledge** and **Broad Intelligence**.	The current industry standard. Spans 57 subjects, including high school and professional topics (e.g., law, computer science, history). Measures academic knowledge and cognitive reasoning ability.	Establishing a *base model's* overall intelligence or benchmarking against SOTA public models.
Big-Bench Hard (BBH)	**Advanced Cognitive Abilities** and **Multi-step Reasoning**.	Focuses on a collection of *challenging* multi-step reasoning tasks where simple pattern matching fails.	Benchmarking a base model's overall intelligence or verifying alignment efforts, particularly for complex tasks.
HELM (Holistic Evaluation of Language Models)	**Transparent, Multi-dimensional Performance Profile**.	Evaluates models across diverse axes, including **accuracy, fairness, robustness, and efficiency**.	Providing a comprehensive, multi-dimensional assessment beyond simple accuracy.

For domain-specific projects, MMLU is ideal for establishing general intelligence, while GLUE is better suited for evaluating performance in specialized NLU fine-tuning.

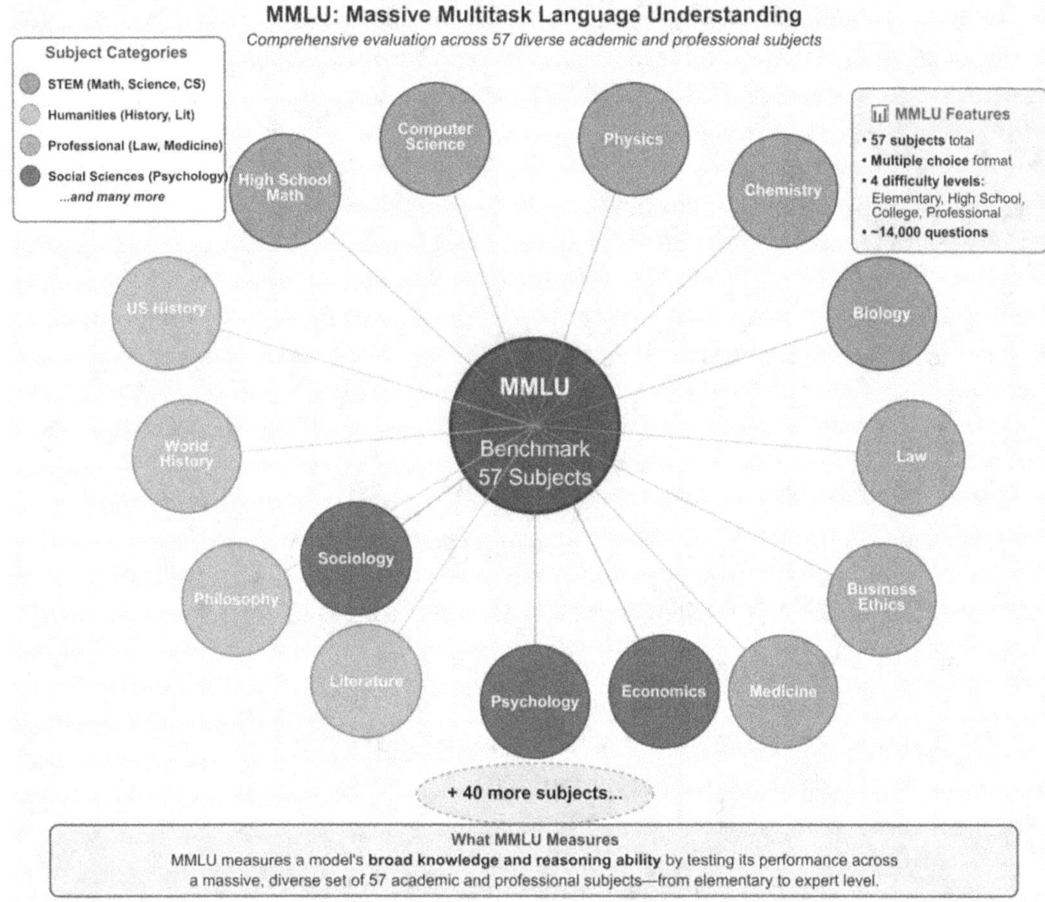

Figure 12-2. *MMLU Benchmark—Multi-Domain Evaluation Framework. Hub-and-spoke diagram of the MMLU benchmark's 57 subject areas organized by category: STEM (blue), Humanities (orange), Professional fields (green), and Social Sciences (purple). MMLU evaluates model knowledge and reasoning across elementary to professional difficulty levels with ~14,000 multiple-choice questions*

Recipe 2: Evaluating a Model on a GLUE Task

Goal: Evaluate a pre-trained model fine-tuned for sentiment analysis on its corresponding GLUE task (SST-2) to verify its performance.

Libraries: evaluate, datasets, transformers, torch

Platform: This recipe runs well in Google Colab.

```
# ## Recipe 2: Evaluating a Model on a GLUE Task
# This recipe loads a model fine-tuned for sentiment analysis
# (the GLUE SST-2 task) and evaluates it on the validation dataset.
# -------------------------------------------------------------------

# -------------------------------------------------------------------
# Recipe 2: GLUE Task Evaluation
# -------------------------------------------------------------------

import torch
from transformers import pipeline
from datasets import load_dataset
import evaluate
from tqdm import tqdm

print("\n--- Recipe 2: Evaluating on GLUE (SST-2) ---")

# 1. Load the GLUE SST-2 (Sentiment) validation dataset
print("Loading GLUE SST-2 dataset...")
dataset = load_dataset("glue", "sst2", split="validation")
# Let's just use the first 200 examples for a quick test
dataset_subset = dataset.select(range(200))
references = dataset_subset["label"]

# 2. Load a pre-trained sentiment pipeline
print("Loading pre-trained sentiment model...")
model_id = "distilbert-base-uncased-finetuned-sst-2-english"
sentiment_classifier = pipeline(
    "text-classification",
    model=model_id,
    device=0 if torch.cuda.is_available() else -1
)
```

```python
# 3. Get predictions for the dataset
print(f"Running predictions on {len(dataset_subset)} samples...")
# The pipeline returns {'label': 'POSITIVE', 'score': 0.99...}
# We need to convert 'POSITIVE' to 1 and 'NEGATIVE' to 0
label_to_int = {"POSITIVE": 1, "NEGATIVE": 0}

sentences = list(dataset_subset["sentence"])

predictions = []
for output in tqdm(sentiment_classifier(sentences)):
    predictions.append(label_to_int[output["label"]])

# 4. Load the GLUE metric and compute
print("Calculating accuracy...")
glue_metric = evaluate.load("glue", "sst2")
accuracy_results = glue_metric.compute(predictions=predictions,
references=references)

print("\n--- GLUE SST-2 Results ---")
print(f"Accuracy on {len(dataset_subset)} samples: {accuracy_
results['accuracy']:.4f}")
```

Recipe 3: Running the MMLU Benchmark

Goal: Run the full MMLU benchmark on a model using a standard, open source framework (Language Model Evaluation Harness).

> **Libraries:** lm-evaluation-harness, transformers, accelerate, bitsandbytes
>
> **Platform:** This recipe is perfect for a Google Colab GPU runtime.

```python
# ## Recipe 3: Running the MMLU Benchmark
#
# This recipe uses the `lm-evaluation-harness` to run the
# MMLU (Massive Multitask Language Understanding) benchmark
# on a model from the Hub.
#
# **This will take a long time (15-30+ minutes)!**
# Recipe 3: MMLU
# ------------------------------------------------------------------------
```

```python
print("\n--- Recipe 3: Running MMLU Benchmark ---")
print("This will take 15-30+ minutes...")

# We use ! to run this as a shell command
# --model hf: Use a Hugging Face model
# --model_args: Load 'google/gemma-2b' in bfloat16
# --tasks mmlu: Run the MMLU benchmark
# --num_fewshot 5: Provide 5 examples in the prompt (standard for MMLU)
# --batch_size auto: Let the harness decide the best batch size
# --output_path: Where to save the detailed results

!lm_eval --model hf \
    --model_args pretrained=google/gemma-2b,dtype=bfloat16 \
    --tasks mmlu \
    --num_fewshot 5 \
    --batch_size auto \
    --output_path ./mmlu_results

print("\nMMLU evaluation complete.")
print("Detailed results are saved in ./mmlu_results/results.json")
print("You can view the summary by running: !cat ./mmlu_results/
results.json")
```

Qualitative Assessment: LLM-As-a-Judge

A limitation of automated metrics is their lack of **human alignment**. A metric may
report a high score, yet a human user may perceive the output as repetitive, unnatural, or
unhelpful.

Human Evaluation

Human evaluation remains the definitive gold standard. It involves presenting two (or
more) anonymous model outputs to human raters and asking them to score the output
based on subjective criteria (e.g., "Which response is more helpful?"). While this yields
the highest quality data, it is inherently slow and expensive to scale.

LLM-As-a-Judge

LLM-as-a-Judge is a modern, scalable approach to qualitative assessment. The methodology is to use a competent, aligned model (e.g., a proprietary or large instruction-tuned model) to evaluate the output of a smaller **candidate model.**

The judge model receives the user's prompt, the candidate model's response, and a detailed **grading rubric**. It is then instructed to provide a numerical score and a **written critique** that justifies the assessment. This procedure is illustrated in Figure 12-3.

- **Advantages:** This approach is faster and cheaper than human annotation while providing rich, qualitative feedback that simple overlap metrics cannot deliver.

- **Considerations:** The judge model itself can introduce bias (e.g., favoring specific output lengths or styles), requiring careful prompt engineering.

This is a powerful technique for rapidly iterating on fine-tuned models where subjective quality is paramount.

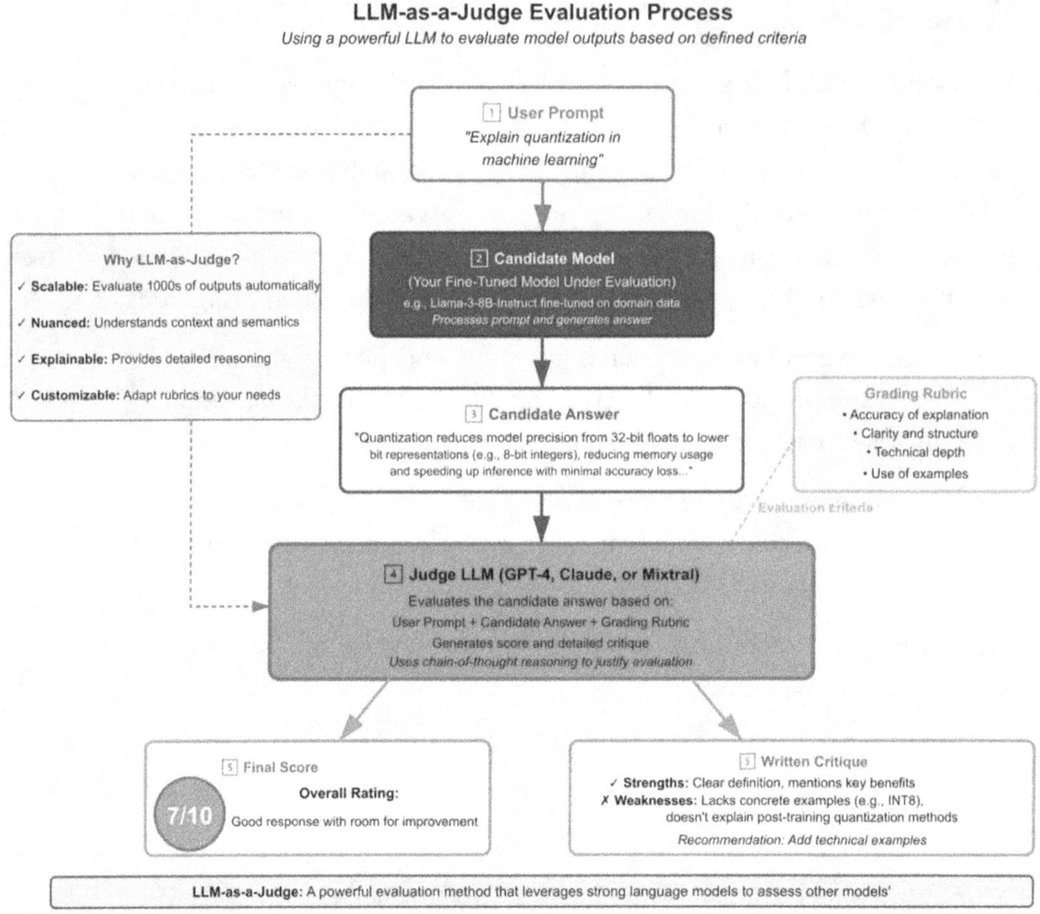

Figure 12-3. *LLM-As-a-Judge Evaluation Pipeline.*
*An automated evaluation workflow where a user prompt (1) feeds a candidate
model (2) to generate an answer (3). The judge LLM (4) evaluates the answer
using the original prompt and grading rubric, producing a numerical score and
detailed critique (5). This approach enables scalable, nuanced model evaluation
with explainable feedback*

Recipe 4: Implementing a Basic LLM-As-a-Judge

Goal: Utilize a powerful instruction-tuned model (e.g., Mistral-7B) to programmatically
assess the quality of an answer generated by a smaller model.

Libraries: transformers, torch, accelerate, bitsandbytes

Platform: This recipe is optimized for Google Colab. **(Refer to the colab_recipes_
for_chapter_12.py file for a runnable example.)**

(The code for this recipe is in the colab_recipes_for_chapter_12.py file.)

```python
# ## Recipe 4: Implementing a Basic LLM-as-a-Judge
#
# This recipe loads a powerful "Judge" model (Mistral-7B-Instruct)
# in 4-bit to fit in Colab. It then uses this judge to
# grade the quality of a pre-written answer.
# ------------------------------------------------------------------

# ------------------------------------------------------------------
# Recipe 4: LLM-as-a-Judge
# ------------------------------------------------------------------

import torch
from transformers import AutoTokenizer, AutoModelForCausalLM, BitsAndBytesConfig

print("\n--- Recipe 4: LLM-as-a-Judge ---")

# 1. Load the "Judge" model in 4-bit
# We use a strong instruction-tuned model
judge_model_id = "mistralai/Mistral-7B-Instruct-v0.1"
print(f"Loading judge model: {judge_model_id} (in 4-bit)...")

bnb_config = BitsAndBytesConfig(
    load_in_4bit=True,
    bnb_4bit_quant_type="nf4",
    bnb_4bit_compute_dtype=torch.bfloat16,
    bnb_4bit_use_double_quant=False,
)
judge_model = AutoModelForCausalLM.from_pretrained(
    judge_model_id,
    quantization_config=bnb_config,
    device_map="auto",
)
judge_tokenizer = AutoTokenizer.from_pretrained(judge_model_id)
if judge_tokenizer.pad_token is None:
    judge_tokenizer.pad_token = judge_tokenizer.eos_token
print("Judge model loaded.")
```

```python
# 2. Define the prompt, reference, and our model's output
user_prompt = "Explain the concept of 'quantization' in large language
models."

reference_answer = (
    "Quantization in LLMs is a technique to reduce the model's memory and "
    "computational cost. It works by converting the model's weights from "
    "high-precision formats (like 32-bit floats) to lower-precision
    formats "
    "(like 8-bit or 4-bit integers). This makes the model smaller and faster "
    "for inference, but can lead to a small loss in accuracy."
)

model_output_to_test = (
    "Quantization is when you make a model. It uses bits and floats. "
    "The 32-bit floats become 4-bit ints. This is good for speed."
)

# 3. Create the "Judge Prompt" (the rubric)
# We use the Mistral instruction format [INST] ... [/INST]
judge_prompt_template = """
[INST]
You are an expert evaluator for large language models. You will be given a
user prompt, a "reference" answer, and a "candidate" answer from a model.
Your task is to rate the "candidate" answer on a scale of 1 to 10 for its
helpfulness, accuracy, and clarity, comparing it to the reference answer.
You must provide your reasoning first, followed by a final score in the
format "Final Score: <score>".

**User Prompt:**
{prompt}

**Reference Answer:**
{reference}

**Candidate Answer:**
{candidate}
[/INST]
```

```
Reasoning:
"""

# 4. Define the judging function
def get_judge_critique(prompt, reference, candidate):
    formatted_prompt = judge_prompt_template.format(
        prompt=prompt,
        reference=reference,
        candidate=candidate
    )

    inputs = judge_tokenizer(formatted_prompt, return_tensors="pt").
    to(judge_model.device)

    print("Judge is thinking...")
    # Generate the critique
    with torch.no_grad():
        outputs = judge_model.generate(
            **inputs,
            max_new_tokens=512,
            pad_token_id=judge_tokenizer.eos_token_id,
            do_sample=False # We want a deterministic critique
        )

    # Decode the response and skip the prompt part
    response_text = judge_tokenizer.decode(outputs[0], skip_special_
    tokens=True)
    critique = response_text[len(formatted_prompt.replace("[INST]", "").
    replace("[/INST]", "")):].strip()
    return critique

# 5. Run the evaluation
critique = get_judge_critique(user_prompt, reference_answer, model_output_
to_test)

print("\n--- LLM-as-a-Judge Critique ---")
print(critique)
```

Comparative Analysis: Selecting the Right Metric

Practical evaluation relies on utilizing the correct tool for the specific task and context. This table compares the utility and application of the core metrics discussed.

Metric	Primary Task(s)	Measurement Focus	Implementation Complexity/Cost	Best Application Context
ROUGE/BLEU	Summarization, Translation	**Token Overlap:** N-gram matching with reference (Recall/Precision).	**Low Cost.** Simple function call via the evaluate library. Runs instantaneously.	Monitoring incremental improvement during fine-tuning for *specific, narrow generative tasks.*
GLUE/SuperGLUE	NLU (e.g., Classification, Entailment)	**Classification Accuracy** and F1 score on understanding tasks.	**Low Cost.** Direct integration with model training/ evaluation frameworks.	Evaluating specialized models fine-tuned for *classification or comprehension tasks.*
MMLU/Big-Bench	General Knowledge, Multi-step Reasoning	**Broad Intelligence:** Academic knowledge and cognitive reasoning ability.	**High Cost.** Requires complex evaluation harnesses (lm-eval) and significant GPU compute time (long runtime).	Benchmarking a *base model's* overall intelligence against SOTA public models or verifying alignment efforts.
LLM-as-a-Judge	Open-Ended Chat, Instruction Following	**Qualitative Subjectivity:** Helpfulness, coherence, safety, and style.	**Medium Cost.** Requires loading a large, secondary "judge" model and careful prompt design.	Assessing *user experience and qualitative performance* where numerical scores fail (e.g., A/B testing two chatbot responses).

The analogy holds: automated tools function as instruments for measurement (like a thermometer), while qualitative assessment serves as the human or expert check for seasoning and overall quality. A robust evaluation plan utilizes a balanced mix of both methodologies.

Chapter Summary

This chapter has provided you with a rigorous, comprehensive methodology for performance assessment in Generative AI, helping you move from casual observation to objective measurement, grounded in four core criteria: Factual Accuracy and Utility, Linguistic Fluency and Coherence, Safety and Alignment, and Input Robustness. To achieve this, you've covered the full spectrum of evaluation techniques, beginning with rapid, low-cost Automated Metrics (like ROUGE and BLEU) and moving to Standardized Benchmarks (such as GLUE/SuperGLUE for NLU, and MMLU/Big-Bench for broad knowledge and reasoning), and concluding with Qualitative Methods like the scalable LLM-as-a-Judge approach to assess subjective attributes. Ultimately, your robust evaluation plan should utilize a balanced mix of these quantitative and qualitative methodologies to ensure your model achieves both high statistical performance and strong human alignment.

PART V

Advanced Applications and Future Directions

Retrieval-Augmented Generation (RAG)

In the previous chapters, we successfully fine-tuned, optimized, deployed, and evaluated our model. We have a highly performant system that serves high-quality text via a production-ready API. However, if we ask, "What happened in the news yesterday?" or "What is our company's new remote work policy?", it will fail.

This is because a standard LLM operates as a **"closed-book" system**. It can only generate responses based on the patterns and facts it learned during its initial training. Its knowledge is vast but **static** and **public**. It lacks a concept of current events (the "knowledge cutoff" problem) and has no access to your specific, private documents (the "domain adaptation" problem).

Welcome to Chapter 13, where we equip our model with an **"open-book" knowledge base. Retrieval-Augmented Generation (RAG)** is a technique that transforms our LLM from a static repository of information into a reasoning engine that can leverage external data. Instead of relying solely on memorized parameters, we enable it to **retrieve** relevant facts from an external source and use them to construct an accurate answer. This single technique simultaneously resolves issues of hallucination, data freshness, and data privacy.

Grounding LLMs: Why RAG?

The core challenge with a stand-alone LLM is that it is ungrounded. Being **"grounded"** in the context of Large Language Models (LLMs) means that the model is provided with and instructed to use **verifiable, external data** to formulate its answer, rather than relying solely on its internal, memorized training data. A standard LLM is an **ungrounded**, "closed-book" system whose knowledge is static and limited to its training data. This causes it to sometimes **hallucinate**, generating responses that are statistically plausible but factually erroneous. **Retrieval-Augmented Generation (RAG)** is a technique for solving this by **grounding** the model in external, verifiable data. RAG transforms the LLM into an "open-book" reasoning engine that retrieves relevant facts from a knowledge base and uses them to construct its final, fact-based answer (Figure 13-1).

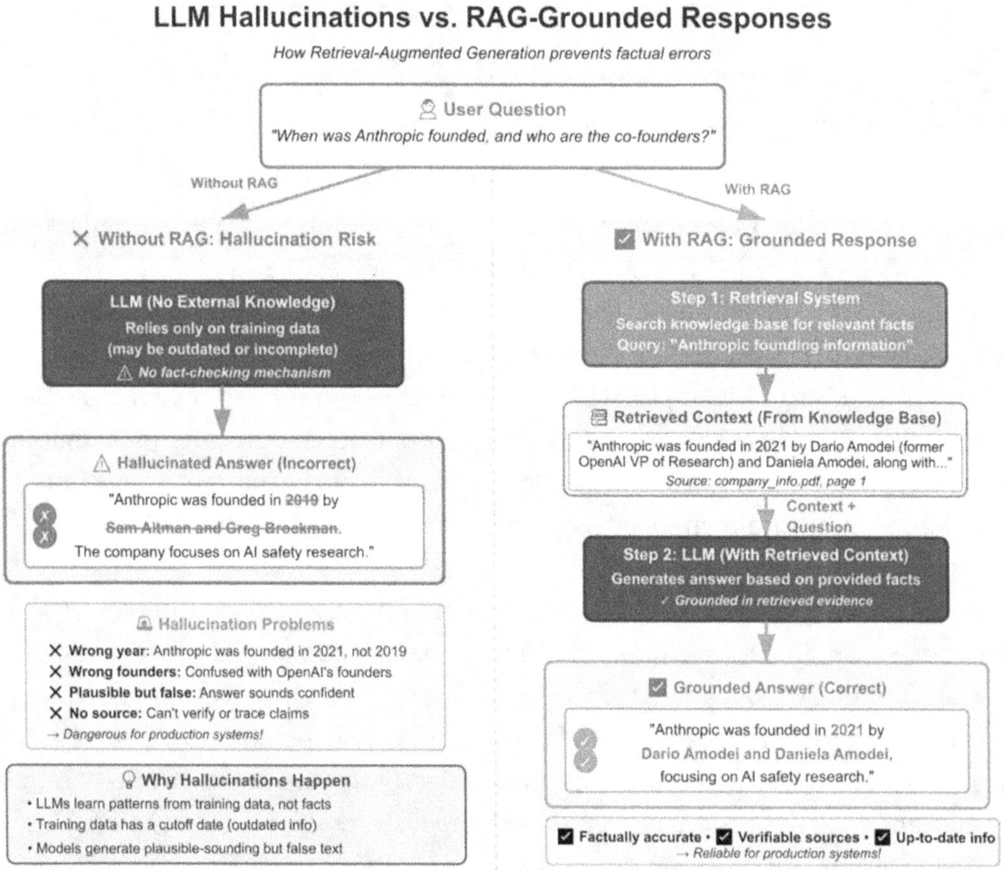

***Figure 13-1.** LLM Hallucinations vs. RAG-Grounded Responses. Comparison showing how LLMs hallucinate factually incorrect information (wrong Anthropic founding year and founders) vs. RAG's two-step process (retrieve relevant documents, then generate grounded answers) that prevents hallucinations through evidence-based generation*

RAG "grounds" the model in verifiable data. Before the LLM generates a response, the system performs a **retrieval step**. The core steps in RAG process are listed below and illustrated in Figure 13-2:

1. **Search:** We query a collection of external documents (e.g., a knowledge base, API, or internal wiki) for information relevant to the user's prompt.

2. **Retrieve:** We extract the most relevant document snippets (the **"context"**).

3. **Augment:** We insert this context into the user's original prompt.

4. **Generate:** We instruct the LLM to answer the user's question *using only the provided context.*

The LLM's role shifts from "answering from memory" to "synthesizing an answer from provided sources." This ensures the outputs are accurate, verifiable, and up to date.

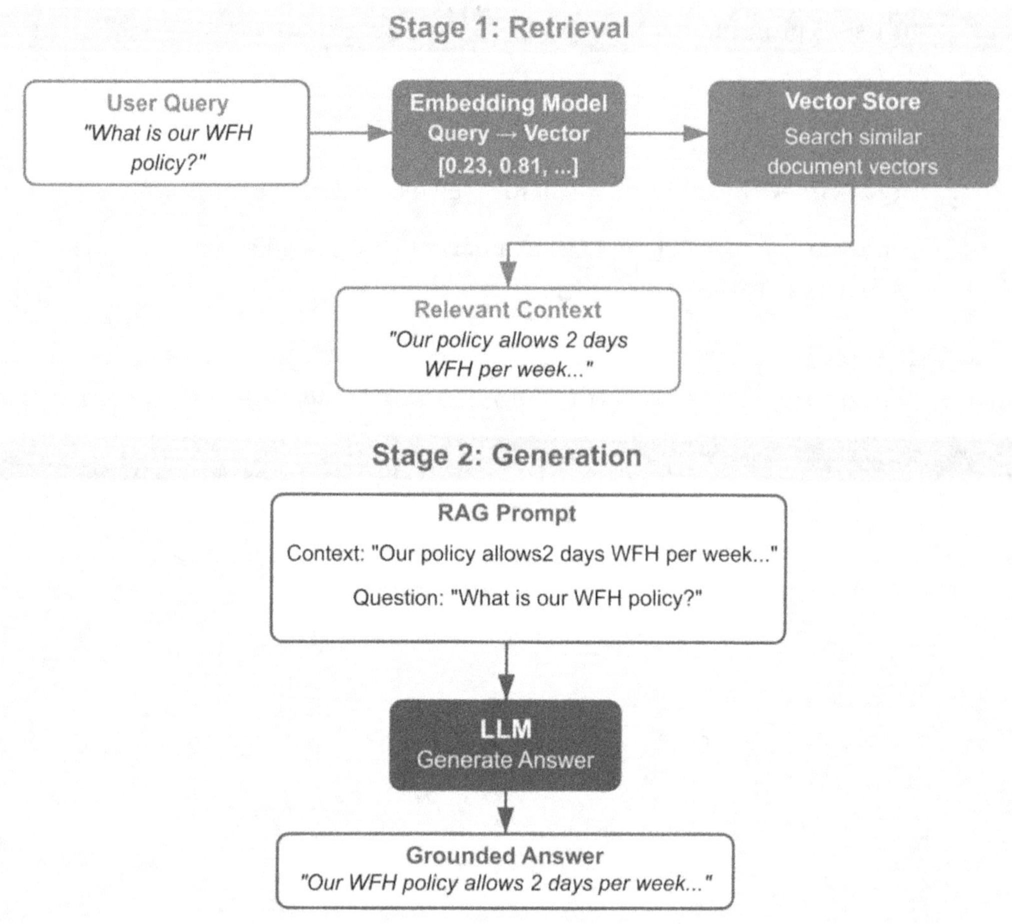

Figure 13-2. *The RAG Process: first, retrieve relevant context from a knowledge base, and then augment the user's prompt with that context to generate a fact-based answer*

Recipe 1: Creating Vector Embeddings

Goal: Convert raw text into high-dimensional numerical vectors (embeddings) using a pre-trained model like SentenceTransformers. This procedure is the fundamental step in semantic search, the core of a Retrieval-Augmented Generation (RAG) pipeline.

This recipe demonstrates the process of transforming human language into a mathematical format, which is essential for grounding a Large Language Model (LLM) in external, verifiable data. The general workflow is

1. **Install and Load Model:** Install the required libraries (e.g., sentence-transformers and torch) and load a small, efficient, pre-trained embedding model (e.g., all-MiniLM-L6-v2).

2. **Define and Process Text:** Define the sentences or text chunks that need to be indexed.

3. **Encode into Vectors:** Use the loaded embedding model's encode() method to convert each piece of text into a high-dimensional vector. These vectors capture the text's semantic meaning.

4. **Inspect Results:** Check the number and shape of the resulting embeddings to verify that the text has been successfully converted into its numerical representation.

What to Expect

This process culminates in a list of dense numerical vectors, where the geometric distance between any two vectors serves as a proxy for the semantic similarity between the original text fragments. When stored in a **Vector Store** (such as FAISS or ChromaDB), these embeddings enable the system to perform a "Similarity Search" to find the most relevant document chunks to a user's question, which is the first crucial step in a RAG pipeline.

```
# Part 1: Install All Dependencies
# ------------------------------------------------------------------------

print("Installing libraries for RAG... This may take a few minutes.")
# For transformers, LLM, and embeddings
!pip install transformers torch accelerate bitsandbytes
!pip install sentence-transformers

# For LangChain
!pip install "langchain==0.1.20" "langchain-community==0.0.34"
!pip install langchain-core pypdf chromadb
!pip install -U bitsandbytes
```

```
# For LlamaIndex
!pip install llama-index llama-index-readers-file llama-index-llms-
huggingface
!pip install faiss-cpu # Using faiss-cpu for easier Colab compatibility

!pip install llama-index-embeddings-huggingface
!pip install llama-index-vector-stores-faiss

print("Installation complete!")

# Part 2: Data Setup
# We will create a dummy HR policy file to use as our external knowledge.
# ----------------------------------------------------------------------

print("Creating dummy document 'hr_policy.txt'...")

hr_policy_text = """
# Fictional Company HR Policy

## 1. Welcome
Welcome to TechCorp! We are excited to have you. This document outlines our
key policies.

## 2. Remote Work Policy
TechCorp offers a hybrid work model. Employees are expected to be in the
office at least 3 days a week.
Full-time remote work (work-from-home) is considered an exception and
requires direct approval from a VP.
All remote employees must maintain a dedicated, quiet workspace and a
stable internet connection.

## 3. Paid Time Off (PTO)
Our PTO policy is flexible. Full-time employees receive 20 days of PTO per
year, which includes vacation and sick leave.
PTO must be approved by your manager at least 2 weeks in advance, except in
cases of emergency.

## 4. Code of Conduct
TechCorp is committed to a professional and inclusive environment.
We have a zero-tolerance policy for harassment of any kind.
```

We also have a strict policy on remote work etiquette. All employees
working from home must be available during core business hours.
"""

```python
with open("hr_policy.txt", "w") as f:
    f.write(hr_policy_text)

print("File 'hr_policy.txt' created successfully.")

# ## Recipe 1: Creating Vector Embeddings
#
# This recipe shows the core concept of embeddings:
# turning text into a list of numbers (a vector).
# ---------------------------------------------------------------------

# ---------------------------------------------------------------------
#
# ---------------------------------------------------------------------

from sentence_transformers import SentenceTransformer

print("\n--- Recipe 1: Creating Vector Embeddings ---")

# 1. Load a pre-trained embedding model
# all-MiniLM-L6-v2 is a small, fast, and high-quality model
print("Loading embedding model: all-MiniLM-L6-v2")
embedding_model = SentenceTransformer('all-MiniLM-L6-v2')

# 2. Define some sentences
sentences = [
    "What is the remote work policy?",
    "How many vacation days do I get?",
    "A developer is writing Python code.",
    "The quick brown fox jumps over the lazy dog."
]

# 3. Encode the sentences into vectors
print("Encoding sentences into vectors...")
embeddings = embedding_model.encode(sentences)

# 4. Inspect the results
```

```
print(f"\nSuccessfully created {len(embeddings)} embeddings.")
print(f"Shape of one embedding (vector): {embeddings[0].shape}")
print(f"The first 5 values of the first vector: {embeddings[0][:5]}")
```

Core Components: The RAG Architecture

A RAG pipeline requires a specialized infrastructure to handle unstructured text data. To build one, we need three key components. The Figure 13-3 depicts the three-phase RAG workflow: Document Indexing, Query & Retrieval, and Answer Generation, which together enable accurate, source-verifiable answers.

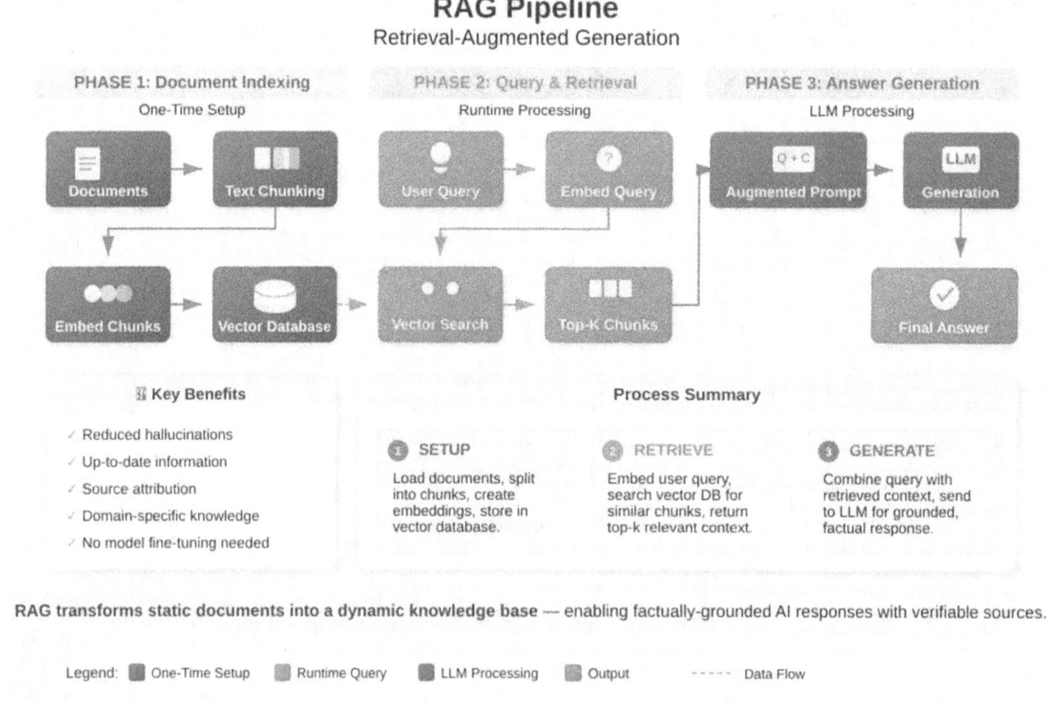

Figure 13-3. *Retrieval-Augmented Generation (RAG) Pipeline Architecture. The diagram illustrates the three-phase RAG workflow: (1) **Document Indexing** (blue), a one-time setup where documents are chunked, embedded into vectors, and stored in a vector database; (2) **Query & Retrieval** (orange), runtime processing where user queries are embedded and matched against stored vectors via similarity search to retrieve the top-k relevant chunks; (3) **Answer Generation** (purple/green), the retrieved context is combined with the original query into an augmented prompt, which the LLM uses to generate a factually grounded response. This architecture enables accurate, source-verifiable answers without requiring model fine-tuning.*

1. Embeddings (Semantic Representation)

Document chunking is a critical first step in the RAG pipeline for preparing unstructured text for semantic search. Since raw, long documents cannot be efficiently stored and searched using simple keyword matching, they are first segmented into smaller, manageable pieces, known as **chunks**. These chunks are typically sized by tokens (e.g., ~500 tokens) and are designed to **overlap** (e.g., by 50 tokens). This overlap is essential because it **preserves semantic context** across chunk boundaries, preventing the loss of meaning when a sentence or concept spans two segments (Figure 13-4).

Once created, an **Embedding Model** converts each text chunk into a high-dimensional numerical **vector**. These vectors capture the semantic meaning of the chunk, enabling the system to efficiently search for similar chunks when a user queries the data.

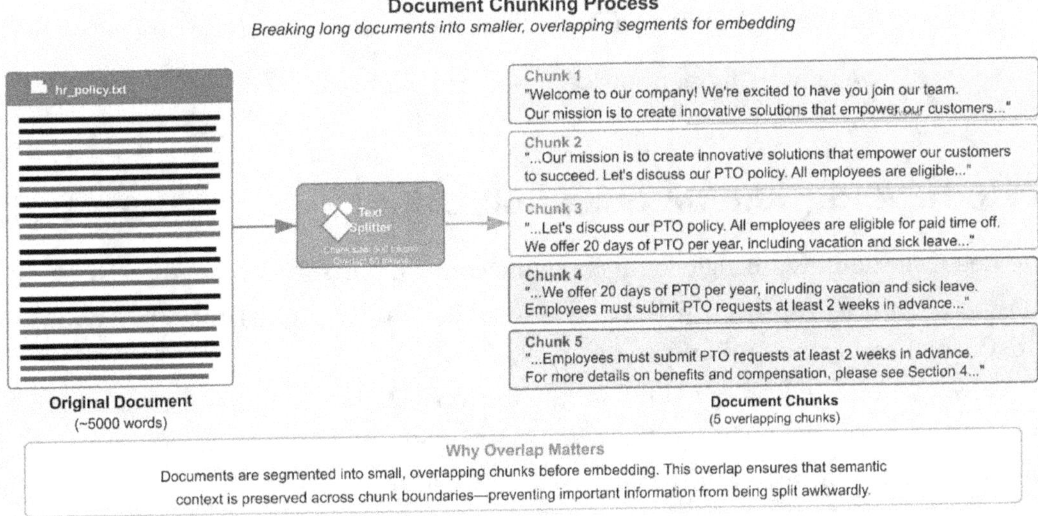

Figure 13-4. Document Chunking Process.Visualization of text segmentation pipeline: raw documents (hr_policy.txt) are split into overlapping chunks (~500 tokens with 50-token overlap) before embedding. The overlap preserves semantic context across chunk boundaries, preventing information loss when sentences or concepts span multiple chunks. Shows five example chunks with highlighted overlapping regions

2. Vector Stores (Indexing and Storage)

A **Vector Store** (or Vector Database) is a specialized database optimized for storing high-dimensional vectors and performing approximate nearest neighbor (ANN) searches at scale:

- **In-Memory/Local:** Ideal for prototyping and smaller datasets

    - **FAISS (Facebook AI Similarity Search):** A highly efficient library for dense vector similarity search

    - **ChromaDB:** A user-friendly, open source vector store that supports local persistence

- **Managed/Cloud:** Designed for production scalability

    - **Pinecone:** A fully managed, cloud-native vector database service

    - **Weaviate:** An open source, modular vector search engine

3. Retrievers (Context Selection Logic)

The **Retriever** defines the logic for fetching the context. The standard method is **Similarity Search (Top-k)**, which retrieves the most similar (e.g., 3) document chunks with the highest cosine similarity to the query.

However, this can lead to redundancy. If the top 3 chunks contain nearly identical information, the context window is utilized inefficiently. A more advanced strategy is **Max Marginal Relevance (MMR)** illustrated in Figure 13-5.

- **MMR** first retrieves a larger set of candidate documents.

- It then iteratively selects documents for the final context by balancing two criteria:

    1. **Relevance:** How similar is the document to the query?

    2. **Diversity:** How dissimilar is the document to the ones *already selected*?

- This results in a context window that covers a broader range of information without redundancy.

Enhancing Retrieval with Re-ranking Strategies

While initial **Similarity Search** (using SentenceTransformers/bi-encoders) is fast and effective for retrieving a large set of *candidate* document chunks, the process alone is often insufficient for production RAG pipelines due to two core issues:

1. **Redundancy (Lack of Diversity):** A simple Top-k search may return highly similar chunks that cover the same information, wasting the LLM's limited context window.

2. **Precision/Coarseness:** Bi-encoders (e.g., SentenceTransformers) compute similarity between two independent vectors (query and chunk). This is fast but less precise than examining the full interaction between the two texts.

To overcome this, a **Re-ranking** step is incorporated between Retrieval and Augmentation:

-   **Cross-Encoders (the Core Technique):** A cross-encoder model takes both the user's query and a candidate document chunk **simultaneously** as input. By comparing the texts together, it generates a far more precise relevance score. Since cross-encoders are computationally intensive, they are only used to "re-rank" the small list of candidates (e.g., the top 20) retrieved by the initial, faster similarity search.

-   **Commercial Re-rankers (e.g., Cohere Re-rank):** These are highly optimized, app-based models that perform the cross-encoding and re-ranking at scale. They provide a final, precisely ordered list of the most relevant and diverse context chunks for the LLM to use.

Incorporating re-ranking ensures that the final context provided to the LLM is not only relevant but also highly precise, significantly boosting the factual accuracy and quality of the final generated response.

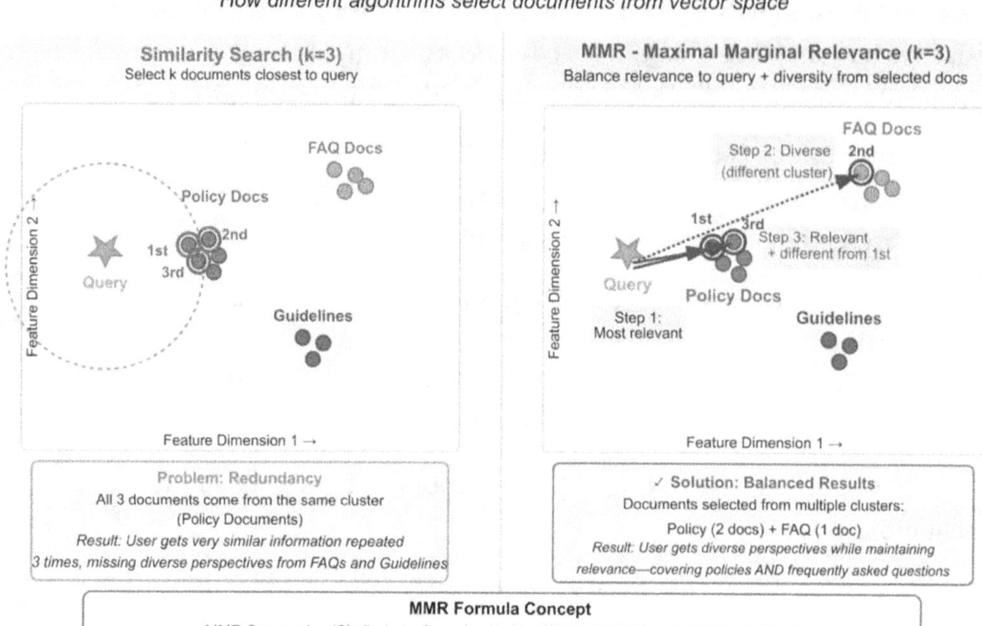

***Figure 13-5.** Similarity Search vs. MMR (Maximal Marginal Relevance). Vector space comparison for query "Physics". Similarity search selects the top 3 closest documents from one cluster (redundant). MMR balances relevance and diversity by selecting from multiple clusters, achieving better information coverage.*

Building RAG Pipelines: LangChain and LlamaIndex

While it is possible to manually code the logic for loading, splitting, embedding, and retrieving data, frameworks like **LangChain** and **LlamaIndex** provide robust abstractions for these tasks:

- **LangChain:** A comprehensive framework for chaining together LLMs with various tools and APIs. It offers a flexible, general-purpose system for building complex applications.

- **LlamaIndex:** A framework optimized explicitly for data ingestion and retrieval. It is designed to be the most efficient interface between your private data and the LLM.

Recipe 2: Simple Q&A System with LangChain and ChromaDB

Goal: Architect an end-to-end Retrieval-Augmented Generation (RAG) pipeline using LangChain for orchestration and an in-memory ChromaDB instance for vector storage to create a fact-based Question & Answer system.

Libraries: langchain, langchain-community, langchain-core, chromadb, pypdf, bitsandbytes, accelerate

This procedure demonstrates the full RAG workflow in action using industry-standard libraries, transforming the components from Recipe 1 (Embeddings) into a functional application. The general workflow is

1. **Load the LLM:** Load a suitable Large Language Model (e.g., google/gemma-2b) using the Hugging Face Pipeline and wrapping it with the LangChain, HuggingFacePipeline adapter.

2. **Load and Split Documents:** Use TextLoader to ingest the source document (e.g., hr_policy.txt). The RecursiveCharacterTextSplitter is then used to break the raw text into manageable, overlapping chunks (e.g., 500 characters with 50 character overlap) to preserve semantic context.

3. **Create Embeddings and Vector Store:** Convert the document chunks into vectors using a model like all-MiniLM-L6-v2. Store these vectors and their corresponding text in the ChromaDB vector store. The store is then exposed as a Retriever.

4. **Define the RAG Chain (Manual):** Construct a Prompt Template that strictly instructs the LLM to use only the provided context to answer the question, mitigating hallucination (e.g., "If the answer is not in the context, say 'I don't know'").

5. **Execute the RAG Flow:** Define a helper function that manually executes the RAG steps: Retrieve (using the vector store to find the top-k relevant chunks for a user's question), Augment (inserting the retrieved context and question into the prompt template), and Generate (calling the LLM to synthesize the final, grounded answer).

What to Expect

This end-to-end system can answer questions that are factually grounded in the external document (*hr_policy.txt*). You will observe how the LLM successfully answers policy-related questions and, importantly, correctly responds with "I don't know" when asked about a topic (like dental plans) that is not covered in the retrieved context, demonstrating the anti-hallucination benefit of RAG.

```python
# Recipe 2: LangChain + ChromaDB (New-style, no RetrievalQA)
# -----------------------------------------------------------------

import torch
from transformers import AutoTokenizer, AutoModelForCausalLM,
BitsAndBytesConfig, pipeline
from langchain_community.llms.huggingface_pipeline import
HuggingFacePipeline

from langchain_community.document_loaders import TextLoader
from langchain_text_splitters import RecursiveCharacterTextSplitter
from langchain_community.embeddings import HuggingFaceEmbeddings
from langchain_community.vectorstores import Chroma

from langchain_core.prompts import PromptTemplate

print("\n--- Recipe 2: Q&A with LangChain & ChromaDB (New-style RAG) ---")

# --- 1. Load the LLM (as a LangChain pipeline) ---
print("Loading LLM (google/gemma-2b)... This will take a moment.")
model_id = "google/gemma-2b"

tokenizer = AutoTokenizer.from_pretrained(model_id)
model = AutoModelForCausalLM.from_pretrained(
    model_id,
```

```python
    torch_dtype=torch.float16,
    device_map="auto",
)

# HF text-generation pipeline
text_gen_pipeline = pipeline(
    "text-generation",
    model=model,
    tokenizer=tokenizer,
    max_new_tokens=200,
    do_sample=True,
    temperature=0.7,
)

# Wrap in LangChain LLM
llm = HuggingFacePipeline(pipeline=text_gen_pipeline)
print("LLM loaded successfully.")

# --- 2. Load and Split Documents ---
print("Loading and splitting document 'hr_policy.txt'...")

loader = TextLoader("hr_policy.txt")
documents = loader.load()

text_splitter = RecursiveCharacterTextSplitter(chunk_size=500, chunk_
overlap=50)
doc_splits = text_splitter.split_documents(documents)

# --- 3. Create Embeddings and Vector Store ---
print("Creating embeddings and ChromaDB vector store...")
embedding_model_name = "all-MiniLM-L6-v2"
embeddings = HuggingFaceEmbeddings(model_name=embedding_model_name)

vectorstore = Chroma.from_documents(documents=doc_splits,
embedding=embeddings)
retriever = vectorstore.as_retriever()
print("Vector store created.")

# --- 4. Define a simple RAG helper (manual chain) ---
```

```python
prompt_template = """
You are an HR policy assistant. Use ONLY the context below to answer the
question.
If the answer is not in the context, say "I don't know" and do not guess.

Context:
{context}

Question:
{question}

Answer:
"""

prompt = PromptTemplate(
    template=prompt_template,
    input_variables=["context", "question"],
)

def build_context(question, k: int = 4) -> str:
    """Retrieve top-k chunks and join them into a context string."""
    # New LangChain retriever API: use .invoke()
    docs = retriever.invoke(question)

    # In most cases, docs is a list of Document objects
    try:
        return "\n\n".join(d.page_content for d in docs)
    except AttributeError:
        # Fallback if some other structure is returned
        return "\n\n".join(str(d) for d in docs)

def answer_question(question: str) -> str:
    """Manual RAG: retrieve -> build prompt -> call LLM."""
    context = build_context(question)
    full_prompt = prompt.format(context=context, question=question)
    # HuggingFacePipeline LLM supports .invoke(prompt_text)
    return llm.invoke(full_prompt)

# --- 5. Ask Questions! ---
```

```python
print("\n--- Asking RAG-powered questions ---")

query_1 = "How many PTO days do I get?"
print(f"Query: {query_1}")
answer_1 = answer_question(query_1)
print(f"Answer: {answer_1}\n")

query_2 = "What is the policy on full-time remote work?"
print(f"Query: {query_2}")
answer_2 = answer_question(query_2)
print(f"Answer: {answer_2}\n")

# This question is NOT in the document
query_3 = "What is the company's dental plan?"
print(f"Query: {query_3}")
answer_3 = answer_question(query_3)
print(f"Answer: {answer_3}\n")
```

Recipe 3: RAG with LlamaIndex and FAISS

Goal: Implement a Retrieval-Augmented Generation (RAG) pipeline using LlamaIndex, which is optimized for data ingestion and retrieval, and persist the high-performance FAISS vector index to disk for efficient reuse across sessions.I

Libraries: llama-index, llama-index-readers-file, llama-index-llms-huggingface, faiss-cpu

This procedure shifts the focus to LlamaIndex, a specialized framework for connecting LLMs to private data, and demonstrates how to handle vector storage persistence—a critical feature for production RAG systems. The general workflow is

1. **Load and Configure Components:** Reload the LLM (e.g., Gemma-2B) and the embedding model (e.g., all-MiniLM-L6-v2), wrapping them in their respective LlamaIndex classes (HuggingFaceLLM and HuggingFaceEmbedding). Set these as the globalSettings for LlamaIndex.

2. **Load Data:** Use LlamaIndex's SimpleDirectoryReader to load the raw documents (hr_policy.txt).

3. **Create FAISS Index:** Initialize a FAISS index (a highly efficient library for vector search). The VectorStoreIndex is then built from the documents, which automatically performs chunking, embedding, and storing the vectors in the FAISS index.

4. **Persist to Disk:** Use the index's storage context to save the entire FAISS index to a local directory. This is the mechanism for persisting the index, allowing it to be loaded instantly in future sessions without the time-consuming re-embedding process.

5. **Create and Query Engine:** Create a query_engine from the index. This engine encapsulates the entire RAG logic, allowing you to submit questions and receive factually grounded answers from the indexed knowledge base.

What to Expect

This process results in a full-featured RAG system that is specialized for private data QA. You will see that the index can be saved and reloaded, eliminating the need to re-build the knowledge base for every use. The system successfully answers fact-based questions like "How many PTO days do employees get?" by retrieving context from the indexed document and generating a grounded response.

```python
# ## Recipe 3: RAG with LlamaIndex & FAISS
#
# This builds a RAG pipeline using LlamaIndex,
# which is specialized for this task. It uses FAISS
# for the vector store and shows how to save it to disk.
# ---------------------------------------------------------------------------

# ---------------------------------------------------------------------------
# Recipe 3: LlamaIndex + FAISS
# ---------------------------------------------------------------------------

import torch
from transformers import AutoTokenizer, AutoModelForCausalLM,
BitsAndBytesConfig
from llama_index.llms.huggingface import HuggingFaceLLM
from llama_index.core import Settings, SimpleDirectoryReader,
VectorStoreIndex
```

```python
from llama_index.embeddings.huggingface import HuggingFaceEmbedding
from llama_index.vector_stores.faiss import FaissVectorStore
import faiss # Import faiss
import os

print("\n--- Recipe 3: RAG with LlamaIndex & FAISS ---")

# --- 1. Load the LLM (LlamaIndex wrapper) ---
# We already loaded the model and tokenizer, but LlamaIndex needs them
# in its own wrapper. We can re-use the ones from the previous cell
# if that cell was run, or reload them here if not.
# For simplicity, this cell is self-contained.

tokenizer_li = AutoTokenizer.from_pretrained(model_id)
model_li = AutoModelForCausalLM.from_pretrained(
    model_id,
    torch_dtype=torch.float16,
    device_map="auto",
)

# LlamaIndex HuggingFaceLLM wrapper
llm_li = HuggingFaceLLM(
    model=model_li,
    tokenizer=tokenizer_li,
    max_new_tokens=200,
    generate_kwargs={"do_sample": True, "temperature": 0.7},
    device_map="auto",
)
print("LLM loaded.")

# --- 2. Configure Settings (Embedding Model) ---
print("Configuring LlamaIndex Settings...")
embedding_model_name = 'all-MiniLM-L6-v2'
embed_model_li = HuggingFaceEmbedding(model_name=embedding_model_name)

# Set the global settings for LlamaIndex
Settings.llm = llm_li
Settings.embed_model = embed_model_li
```

```python
# --- 3. Load Data ---
print("Loading documents with SimpleDirectoryReader...")
# This can read from a directory, but we'll point it to our file
reader = SimpleDirectoryReader(input_files=["hr_policy.txt"])
documents = reader.load_data()

# --- 4. Create FAISS Vector Store and Index ---
print("Creating FAISS vector store...")
# Setup a FAISS vector store
d = 384  # Dimensions of all-MiniLM-L6-v2
faiss_index = faiss.IndexFlatL2(d)
vector_store_li = FaissVectorStore(faiss_index=faiss_index)

# Build the VectorStoreIndex, which will chunk, embed, and store
# This combines the "VectorStore" and "Index" steps
from llama_index.core.storage.storage_context import StorageContext
storage_context = StorageContext.from_defaults(vector_store=vector_
store_li)
index = VectorStoreIndex.from_documents(
    documents, storage_context=storage_context
)
print("FAISS Index built.")

# --- 5. Save the Index to Disk ---
if not os.path.exists("./faiss_index"):
    os.makedirs("./faiss_index")
print("Persisting index to disk at './faiss_index'...")
index.storage_context.persist(persist_dir="./faiss_index")
print("Index saved. You can load this next time instead of rebuilding.")

# --- 6. Create Query Engine and Ask Questions ---
print("Creating query engine...")
query_engine = index.as_query_engine()

print("\n--- Asking LlamaIndex-powered questions ---")
query_1 = "What is the code of conduct regarding harassment?"
print(f"Query: {query_1}")
response_1 = query_engine.query(query_1)
```

```
print(f"Answer: {response_1}\n")

query_2 = "How many PTO days do employees get?"
print(f"Query: {query_2}")
response_2 = query_engine.query(query_2)
print(f"Answer: {response_2}\n")
```

Recipe 4: Comparing Retrieval Strategies (MMR)

Goal: Demonstrate the performance difference between standard "Similarity Search" and "Max Marginal Relevance (MMR)" when handling queries that require diverse information.

Libraries: langchain, chromadb, sentence-transformers

This procedure is designed to highlight a critical refinement in the RAG retrieval step: moving beyond basic relevance to explicitly include **diversity**. It shows how different retrieval algorithms affect the quality of the context passed to the LLM, particularly when a user's question has multiple, distinct areas of relevant information within the knowledge base. The comparison workflow is

1. **Reuse Vector Store:** The experiment reuses the vector store (e.g., ChromaDB) created in **Recipe 2**, which contains the vectorized document chunks.

2. **Define Similarity Retriever:** A standard retriever is created using the "similarity" search type, which returns only the top-k document chunks closest in vector space to the query.

3. **Define MMR Retriever:** A second retriever is created using the more advanced "mmr" (Max Marginal Relevance) search type, which balances two criteria: high relevance to the query AND low similarity to the already selected documents.

4. **Execute and Compare:** A deliberately broad query (e.g., "What are the company's policies on remote work?") is run against both retrievers to show the difference in the resulting set of document chunks.

What to Expect

You will observe that the **Similarity Search** is likely to return multiple chunks that are highly similar to each other, for example, two different paragraphs discussing the "hybrid work model" policy. In contrast, the **MMR Search** is designed to retrieve one chunk about the "hybrid work model" (high relevance) and a different, non-redundant chunk about the "Code of Conduct" for remote work (maintaining diversity). This comparison demonstrates that MMR efficiently utilizes the LLM's context window by providing a broader range of necessary information.

```python
# ## Recipe 4: Comparing Retrieval Strategies (MMR)
#
# This recipe uses the LangChain vector store from Recipe 2
# to show the difference between "Similarity Search" and
# "Max Marginal Relevance (MMR)" search.
# -------------------------------------------------------------------

# -------------------------------------------------------------------
# Recipe 4: MMR vs. Similarity Search
# -------------------------------------------------------------------

print("\n--- Recipe 4: Comparing Retrieval Strategies ---")

# We re-use the 'vectorstore' object from Recipe 2
# If you didn't run Recipe 2, this will fail.
# Ensure Recipe 2 (LangChain + ChromaDB) cell has been run.

if 'vectorstore' not in locals():
    print("Error: 'vectorstore' not found.")
    print("Please run Recipe 2 (LangChain + ChromaDB) cell first.")
else:
    print("Using vector store from Recipe 2.")

    # This query is broad and matches two distinct policy areas:
    # 1. The remote work "exception" policy
    # 2. The remote work "etiquette" policy
    query = "What are the company's policies on remote work?"
    print(f"\nQuery: {query}\n")

    # --- 1. Standard Similarity Search ---
```

```python
print("--- 1. Similarity Search (k=2) ---")
# This will just find the 2 closest chunks
retriever_sim = vectorstore.as_retriever(search_type="similarity",
search_kwargs={"k": 2})
results_sim = retriever_sim.invoke(query)

for i, doc in enumerate(results_sim):
    print(f"Result {i+1} (Similarity):\n{doc.page_content}\n")

# --- 2. Max Marginal Relevance (MMR) Search ---
print("--- 2. Max Marginal Relevance (MMR) (k=2) ---")
# This will find 2 chunks that are both relevant AND diverse
retriever_mmr = vectorstore.as_retriever(search_type="mmr", search_
kwargs={"k": 2})
results_mmr = retriever_mmr.invoke(query)

for i, doc in enumerate(results_mmr):
    print(f"Result {i+1} (MMR):\n{doc.page_content}\n")

print("--- Comparison ---")
print("Notice how 'Similarity' might return two very similar chunks
about the hybrid model.")
print("MMR is more likely to return one chunk about the hybrid model
AND one chunk about the 'Code of Conduct' for remote work.")
```

Practical Limitations of RAG

While Retrieval-Augmented Generation (RAG) is a powerful technique, it is not without its limitations, which must be managed in a production system:

- **Context Window Limits:** The maximum length of text the LLM can process (the context window) is finite. Even with RAG, only a limited number of retrieved document chunks can be passed to the model, meaning extremely complex or broad questions may exceed this limit.

- **Prompt Injection Risks:** Like other LLM applications, RAG is susceptible to prompt injection, where malicious or misleading instructions are inserted into the retrieved documents or the user query to manipulate the LLM's final response.

- **Retrieval ≠ Truth Guarantee:** The retrieval process only provides the LLM with information from the knowledge base. If the source documents are factually incorrect, outdated, or misleading, the LLM will generate a grounded answer based on those flawed sources.

- **Embedding Drift Over Time:** As the world changes and new data is added, the original vector embeddings may become less relevant or aligned with newer information, potentially degrading retrieval performance over long periods.

- **Chunk-Size Trade-Offs:** Selecting the optimal chunk size is a balancing act. Small chunks may lack necessary context, while large chunks may dilute the relevance signal, making it difficult for the LLM to identify the precise answer.

Chapter Summary

In this chapter, we integrated a critical capability into our LLM system: external knowledge retrieval. We established that **Retrieval-Augmented Generation (RAG)** is the definitive solution for grounding models, mitigating hallucinations, and enabling access to private or real-time data. We examined the structural components of a RAG pipeline: **Embeddings** (semantic representations), Vector Stores (high-performance indexing systems such as **ChromaDB** and **FAISS**), and **Retrieval Logic**. We also used industry-standard frameworks such as **LangChain** and **LlamaIndex** to orchestrate these components. Finally, we executed four complete recipes, ranging from basic embedding generation to building full Q&A systems and implementing advanced retrieval strategies, such as **MMR**.

We also used industry-standard frameworks such as **LangChain** and **LlamaIndex** to orchestrate these components. Finally, we executed four complete recipes, ranging from basic embedding generation to building full Q&A systems, implementing FAISS persistence, and comparing retrieval strategies. We also discussed the **Practical Limitations of RAG**, including context window limits, prompt injection risks, and the issue of flawed source data. Up to this point, our system has processed only textual data. In the next chapter, we will expand our scope to **Multimodal Models**, introducing architectures capable of processing and reasoning about visual data alongside text.

Exploring Vision-Language Models

In previous chapters, we optimized a model proficient in a single modality: **text**. We have fine-tuned, quantized, deployed, evaluated, and augmented it with external knowledge (RAG). However, real-world data is inherently multimodal. Human cognition processes visual, auditory, and textual information simultaneously to build a comprehensive understanding of the environment.

This chapter focuses on integrating visual perception capabilities, transitioning from Large Language Models (LLMs) to **Vision-Language Models (VLMs)**. These multimodal architectures can simultaneously ingest, process, and reason about both image and text data. This capability enables a system to analyze visual inputs, such as a photograph of a refrigerator's contents, and generate contextually relevant text outputs, including recipe recommendations.

Beyond Text: Introduction to Vision-Language Models (VLMs)

A VLM integrates two distinct neural network components into a unified inference engine:

1. **Visual Encoding Module (Vision Encoder):** This component (typically a **ViT** (Vision Transformer) or a **CLIP** encoder) is trained to analyze an image and compress it into a set of "visual embeddings," high-dimensional vectors representing the objects, textures, and spatial relationships within the image.

2. **Reasoning Engine (Language Model):** This is the standard LLM architecture (such as Llama, Gemma, or Mistral). It excels at logical reasoning and text generation but processes only text embeddings natively.

341

B. K. Bolla et al., *Large Language Model Recipes*, https://doi.org/10.1007/979-8-8688-2607-8_14

The critical component of a VLM is the **Projector** (or Adapter) that bridges these two modules. This neural network layer learns to map the *encoder's visual* embeddings into the vector space of the *text* embeddings. Effectively, it translates visual data into "visual tokens" that the LLM can process as if they were part of a textual prompt. Figure 14-1 illustrates the core components of the Vision-Language Model (VLM) architecture, detailing how the visual and language modules are bridged by the projector layer to enable multimodal understanding.

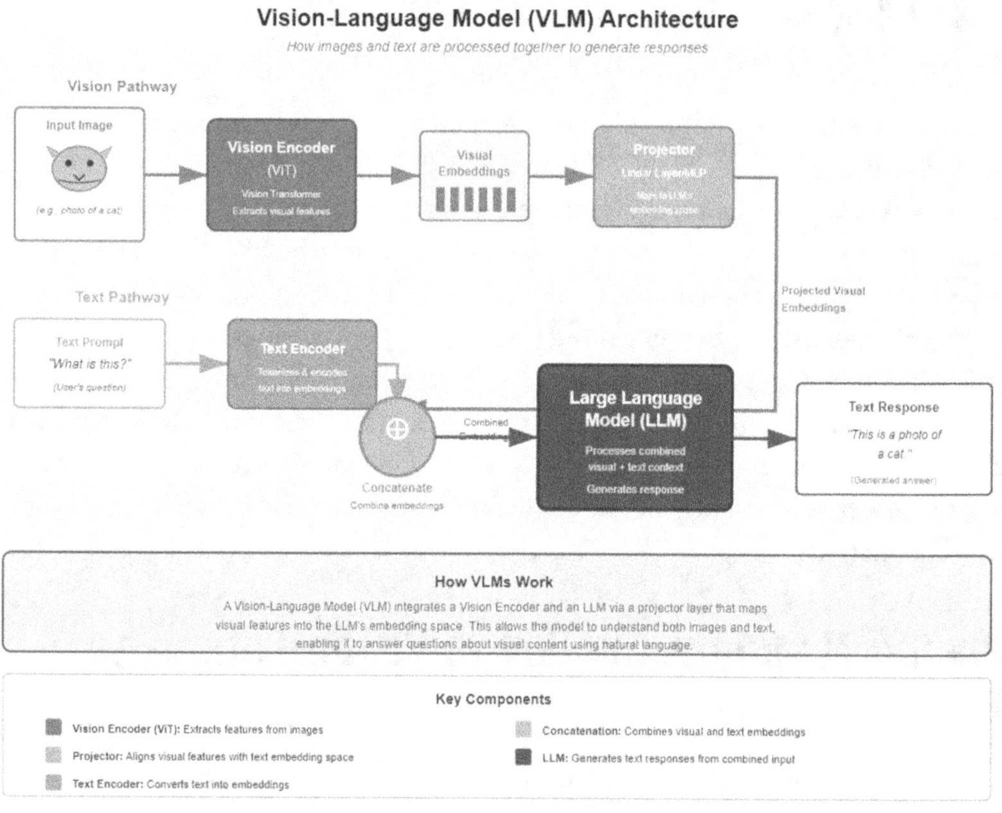

Figure 14-1. *Vision-Language Model (VLM) Architecture.*
Data flow through VLM components: the input image passes through the Vision Encoder (ViT) to generate visual embeddings, which are mapped to the LLM embedding space via the projector layer. Text prompt is encoded separately. Both embeddings are concatenated and fed to the LLM for response generation, enabling multimodal understanding

Key Architectures and Open Models

While numerous VLM architectures exist, most adhere to a few foundational design patterns.

CLIP (Contrastive Language-Image Pre-training)

The CLIP (Contrastive Language-Image Pre-Training) model is a foundational architecture in Vision-Language Models (VLMs), developed by OpenAI. CLIP is not a generative model, but it serves as the foundational embedding layer for most modern VLMs. Developed by OpenAI, it is primarily built on the principle of cross-modal alignment, meaning its primary goal is to learn how to represent images and text in a shared, high-dimensional vector space (an embedding space).

Here are the key aspects of its building and function:

- **Training Data:** It is trained on billions of image-text pairs.

- **Alignment Objective:** During training, CLIP learns to generate embeddings (vectors) such that the vector for an image is **geometrically close** to the vector for its correct textual description and far away from the vectors of incorrect descriptions. This contrastive learning process ensures a strong connection between what an image depicts and what the corresponding text describes.

- **Role in VLMs:** While CLIP itself is not a generative model, it serves as the foundational embedding layer for most modern generative VLMs.

- **Key Capability:** Its strong semantic alignment enables **Zero-Shot Classification**. This means it can classify an image against a set of arbitrary, user-defined text labels (e.g., "dog," "cat," and "car") without being explicitly trained on those categories. It does this by comparing the image's embedding with the embeddings of the potential text labels and selecting the closest match. We will demonstrate this in Procedure 3.

BLIP (Bootstrapping Language-Image Pre-training)

BLIP (and its successor, BLIP-2) from Salesforce is a family of models designed for generative multimodal tasks. Unlike CLIP, which focuses on cross-modal alignment, BLIP is trained explicitly for **conditional generation**.

Here are the key aspects of its building and function:

- **Focus:** Generative multimodal tasks. Unlike CLIP, it focuses on **conditional generation**.

- **Capability:** This allows it to process an image and generate descriptive text.

- **Applications:** It is highly effective for tasks such as

  - **Automated Image Captioning:** Generating a concise textual description of an image's content (demonstrated in Procedure 1)

  - Visual Question Answering (VQA) (implied by its generative focus and the other models' use cases)

LLaVA (Large Language and Vision Assistant)

LLaVA represents a state-of-the-art approach in open source Vision-Language Models (VLMs). Its architecture is specifically designed for efficiency and conversational capability, enabling it to serve as a robust conversational agent.

Here are the key aspects of its building and function:

- **Architecture Components:** LLaVA integrates pre-trained components to optimize its design:

  - **Vision Encoder:** Utilizes a pre-trained **CLIP** encoder (specifically ViT-L/14) for visual perception

  - **Language Model (Reasoning Engine):** Utilizes a powerful, pre-trained **LLM** (e.g., Vicuna or Llama)

- **Integration:** It connects the Vision Encoder and the Language Model via a simple linear **Projector**.

- **Training Strategy:** It employs a highly efficient training method: it fine-tunes *only* the projector and the LLM on a dataset of multimodal instructions, keeping the original Vision Encoder weights frozen.

- **Key Capability:** The result is a robust conversational agent capable of analyzing user-uploaded images and engaging in multi-turn dialogue. It is typically used for complex tasks such as **Visual Question Answering (VQA).**

Implementation Procedures

The following procedures demonstrate the implementation of core multimodal workflows, ranging from basic image description to complex reasoning and zero-shot classification.

Recipe 1: Automated Image Captioning with BLIP

Goal: Generate a concise textual description of an image's content. This task demonstrates the model's ability to translate visual features into natural language.

Libraries: transformers, torch, Pillow

This procedure is designed to showcase the core generative capability of Vision-Language Models, specifically using the BLIP (Bootstrapping Language-Image Pre-Training) architecture for conditional image captioning. It illustrates how a VLM processes an image and generates a descriptive text output.

The captioning workflow is

1. **Dependency Setup:** All required libraries (transformers for the model, Pillow for image handling, etc.) are installed, and necessary sample images are downloaded.

2. **Pipeline Loading:** The image-to-text pipeline is initialized using a pre-trained BLIP model (e.g., Salesforce/blip-image-captioning-large).

3. **Image Input:** A sample image (e.g., sample_cat_on_sofa.jpg) is loaded.

4. **Caption Generation:** The loaded image is passed to the
    captioning pipeline, which utilizes the BLIP model to generate
    a descriptive caption. Parameters such as max_new_tokens are
    used to control the output's length and conciseness.

5. **Output:** The resulting textual caption is printed.

What to Expect

You will observe a generated sentence that accurately and concisely describes
the content of the sample image. This demonstrates the BLIP model's strong focus on
conditional generation and its effectiveness for automated image description tasks.

```python
# Part 1: Install All Dependencies
# -----------------------------------------------------------------------

print("Installing libraries for VLM... This may take a few minutes.")
# For transformers, LLM, and multimodal pipelines
!pip install transformers torch accelerate bitsandbytes
!pip install pillow # For loading images
!pip install requests # For downloading sample images
print("Installation complete!")

# Part 2: Data Setup
# We will download two sample images to use with our recipes.
# -----------------------------------------------------------------------

import requests
from PIL import Image
from io import BytesIO
import os

print("Downloading sample images...")

# Sample 1: A photo of a bus
img_url_1 = "http://images.cocodataset.org/val2017/000000039769.jpg"
img_path_1 = "sample_cat_on_sofa.jpg"

def download_image(url, path):
    try:
        response = requests.get(url)
```

```python
        response.raise_for_status()
        img = Image.open(BytesIO(response.content))
        img.save(path)
        print(f"Saved '{path}'")
    except Exception as e:
        print(f"Failed to download {url}. Error: {e}")

if not os.path.exists(img_path_1):
    download_image(img_url_1, img_path_1)

print("Sample images are ready.")

from transformers import pipeline
from PIL import Image

print("\n--- Recipe 1: Image Captioning (BLIP) ---")

# 1. Load the "image-to-text" pipeline with a BLIP model
print("Loading captioning model: Salesforce/blip-image-captioning-large")
captioner = pipeline(
    "image-to-text",
    model="Salesforce/blip-image-captioning-large",
    device=0 # Use 0 for GPU
)

# 2. Open the sample image
img = Image.open("sample_cat_on_sofa.jpg")

# 3. Generate the caption
print("Generating caption for 'sample_cat_on_sofa.jpg'...")
# We use max_new_tokens to keep the caption concise
caption_result = captioner(img, max_new_tokens=50)

# 4. Print the result
print("\n--- Generated Caption ---")

print(caption_result[0]['generated_text'])
```

Recipe 2: Visual Question Answering (VQA) with LLaVA

Goal: Query the model with specific questions regarding an image to evaluate its multimodal reasoning capabilities.

Libraries: transformers, torch, accelerate, bitsandbytes, Pillow

This procedure demonstrates how a modern, conversational Vision-Language Model like LLaVA (Large Language and Vision Assistant) performs complex, grounded reasoning by integrating visual input with a specific textual question. To handle resource constraints, it often uses a 4-bit quantized version of the model.

The VQA workflow is:

1. **Model Loading:** The LLaVA model and its associated processor are loaded, often utilizing 4-bit quantization (e.g., using a BitsAndBytesConfig).

2. **Input Definition:** A sample image and a specific question about that image are defined.

3. **Prompt Formatting:** The image and question are combined into a structured, specific chat prompt format (e.g., USER: <image>\ n{question}\nASSISTANT:) as required by the LLaVA architecture.

4. **Inference and Generation:** The formatted prompt and image are passed to the model, which generates a natural language response based on its visual and textual understanding.

5. **Output:** The generated text-based answer to the question is printed.

What to Expect

You will observe a concise, factual answer that is grounded in the image content. This demonstrates the model's ability to align visual features (like the color of a shirt) with the semantic query intent (the question about the shirt's color), showcasing its high-level multimodal reasoning. Figure 14-2 illustrates the Visual Question Answering (VQA) Task workflow, detailing the process of integrating visual input with a specific textual question to generate a grounded, factual answer.

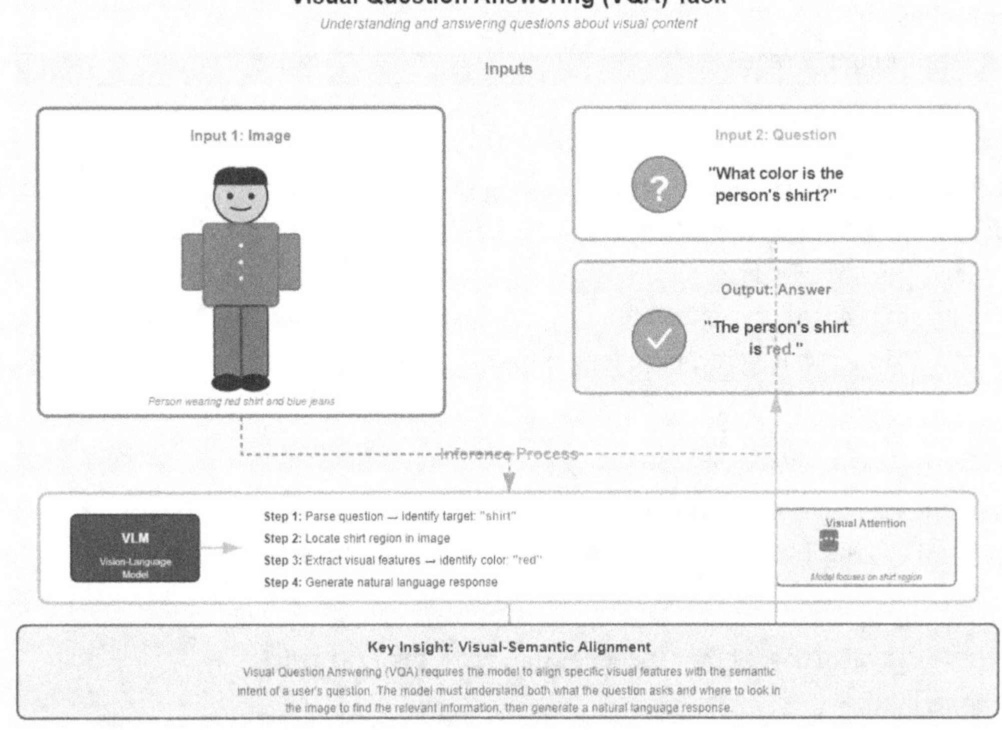

Figure 14-2. *Visual Question Answering (VQA) Task.*
VQA workflow showing inputs (image of person in red shirt and blue jeans + question "What color is the person's shirt?"), inference process with visual attention mechanism highlighting the shirt region, and grounded output ("The person's shirt is red"). Demonstrates alignment of visual features with semantic query intent

```
# ## Recipe 2: Visual Question Answering (LLaVA)
#
# This recipe uses a 4-bit quantized LLaVA model
# to answer a *specific* question about an image.
# This will take a few minutes to load.
# --------------------------------------------------------------------

# --------------------------------------------------------------------
# Recipe 2: VQA with LLaVA (4-bit)
# --------------------------------------------------------------------

import torch
```

```python
from transformers import LlavaForConditionalGeneration, BitsAndBytesConfig,
AutoProcessor
from PIL import Image

print("\n--- Recipe 2: VQA with LLaVA (4-bit) ---")

# 1. Load the 4-bit quantization config
bnb_config = BitsAndBytesConfig(
    load_in_4bit=True,
    bnb_4bit_quant_type="nf4",
    bnb_4bit_compute_dtype=torch.bfloat16,
)

# 2. Load the LLaVA model and processor
model_id = "llava-hf/llava-1.5-7b-hf"
print(f"Loading model: {model_id} (in 4-bit)... This may take a few
minutes.")

model = LlavaForConditionalGeneration.from_pretrained(
    model_id,
    quantization_config=bnb_config,
    device_map="auto",
)
processor = AutoProcessor.from_pretrained(model_id)
print("Model and processor loaded.")

# 3. Define the image and the prompt
# LLaVA requires a specific chat format
img = Image.open("sample_cat_on_sofa.jpg")
question = "What are the cats doing?"
prompt = f"USER: <image>\n{question}\nASSISTANT:"

print(f"\nImage: cat_on_sofa.jpg")
print(f"Question: {question}")

# 4. Process inputs and generate
inputs = processor(text=prompt, images=img, return_tensors="pt").to("cuda")

print("LLaVA is thinking...")
```

```
with torch.no_grad():
    outputs = model.generate(**inputs, max_new_tokens=100)

# 5. Decode and print the response
response_full = processor.decode(outputs[0], skip_special_tokens=True)

# Clean up the output to only show the assistant's answer
response_answer = response_full.split("ASSISTANT:")[-1].strip()
print("\n--- Generated Answer ---")
print(response_answer)
```

Zero-Shot Image Classification Using CLIP

The CLIP (Contrastive Language-Image Pre-Training) model achieves Zero-Shot Classification by leveraging its strong semantic alignment to represent both images and text in a shared high-dimensional embedding space. As illustrated in Figure 14-3, the process begins by encoding the input image (e.g., a bus photo) and candidate text labels (e.g., "car," "truck," and "bus") via the Visual Encoder into separate vectors. The model then calculates the Cosine Similarity between the image vector and each text-label vector to measure their geometric closeness in this shared space. Finally, a Softmax function is applied to these similarity scores to produce probabilities. The label with the highest probability (e.g., 0.95 for "bus") is selected as the correct classification, demonstrating that the model can classify an image against arbitrary text labels without being explicitly trained on those specific categories.

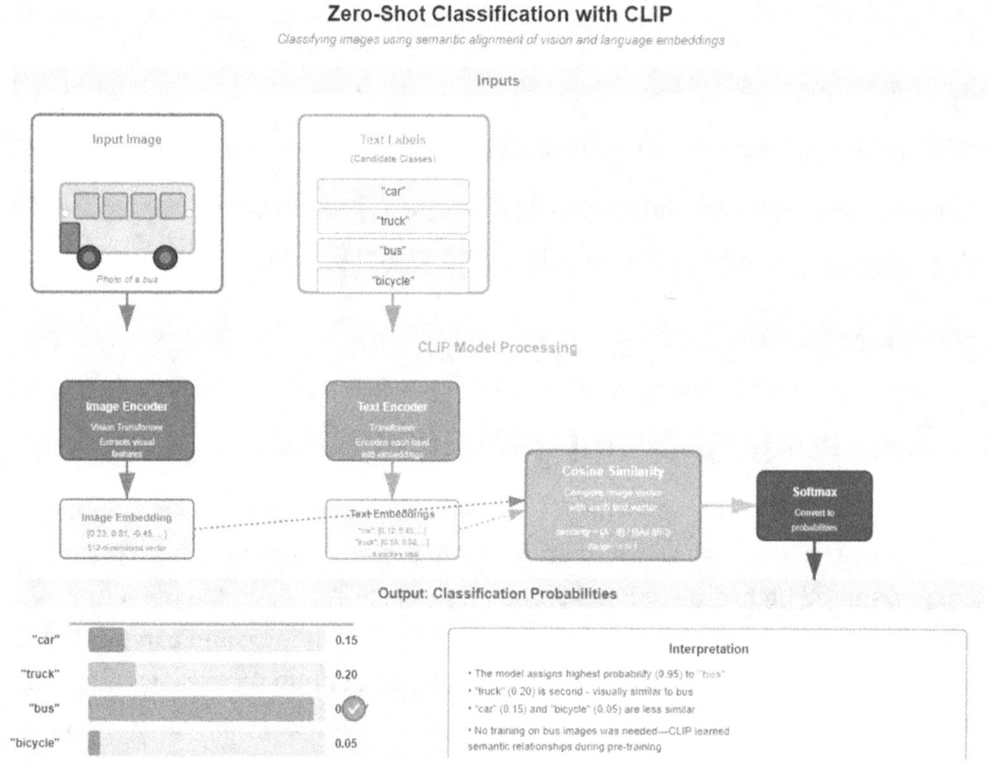

Figure 14-3. *Zero-Shot Classification with CLIP.*
CLIP's embedding alignment process: input image (bus photo) encoded to a vector, candidate text labels ("car", "truck", "bus", "bicycle") encoded separately. Cosine similarity is computed between the image and each text embedding. Softmax converts similarities to probabilities, correctly identifying "bus" (0.95) without task-specific training

Figure 14-3 details the following steps:

1. **Input Encoding:** An input image (represented by a "bus photo" in the figure) is processed by the Visual Encoder and encoded into a high-dimensional **vector** (image embedding).

2. **Label Encoding:** Candidate text labels (e.g., "car," "truck, "bus," and "bicycle") are also encoded as separate **vectors** (text embeddings).

3. **Similarity Calculation:** The model computes the **Cosine Similarity** between the image vector and *each* candidate text-label vector. This measures the geometric closeness in the shared embedding space.

4. **Prediction:** A **Softmax** function is applied to these similarity scores, yielding probabilities.

5. **Result:** The process correctly identifies the image as a "bus" with a high probability (0.95) by finding the closest match between the image and text embeddings.

This workflow showcases how CLIP achieves **Zero-Shot Classification** by leveraging its strong semantic alignment, which connects visual content to textual concepts.

Recipe 3: Zero-Shot Image Classification Using CLIP

Goal: Classify an image against a set of arbitrary, user-defined text labels without any fine-tuning on those specific categories.

Libraries: transformers, torch, Pillow

This procedure uses the foundational CLIP (Contrastive Language-Image Pre-Training) model to demonstrate its core capability: Zero-Shot Classification. It relies on the model's ability to represent images and text in a shared embedding space, demonstrating that a model can understand a concept from its textual description alone, without seeing a labeled image of it during training.

The classification workflow is

1. **Pipeline Loading:** The zero-shot-image-classification pipeline is initialized with a pre-trained CLIP model (e.g., openai/clip-vit-large-patch14).

2. **Input Definition:** A sample image and a list of arbitrary candidate_labels (text descriptions) are defined.

3. **Classification:** The image and the list of text labels are passed to the classifier. The model calculates the similarity (e.g., cosine similarity) between the image's embedding and the embedding of each text label.

4. **Output:** The results, including the score (probability) for each label, are printed, highlighting the most likely match.

What to Expect

You will observe that the model correctly identifies the image content (e.g., "a photo of a cat on a sofa") from a diverse list of choices (e.g., "a drawing of a city"). This demonstrates the power of the CLIP architecture's learned semantic alignment, enabling instant classification without requiring any task-specific retraining.

```python
# ## Recipe 3: Zero-Shot Image Classification (CLIP)
#
# This recipe uses CLIP to classify an image from a
# list of text labels, without any fine-tuning.
# -----------------------------------------------------------------------
# -----------------------------------------------------------------------
# Recipe 3: Zero-Shot Classification with CLIP
# -----------------------------------------------------------------------

from transformers import pipeline
from PIL import Image

print("\n--- Recipe 3: Zero-Shot Classification (CLIP) ---")

# 1. Load the "zero-shot-image-classification" pipeline with CLIP
print("Loading classification model: openai/clip-vit-large-patch14")
classifier = pipeline(
    "zero-shot-image-classification",
    model="openai/clip-vit-large-patch14",
    device=0 # Use 0 for GPU
)

# 2. Open the image and define our candidate labels
img = Image.open("sample_cat_on_sofa.jpg")
candidate_labels = [
    "a photo of a cat on a sofa",
    "a drawing of a city",
    "two astronauts walking on the moon",
    "a man playing a guitar"
]
```

```python
print(f"\nImage: sample_cat_on_sofa.jpg")
print(f"Labels: {candidate_labels}")

# 3. Classify the image
print("Classifying image...")
results = classifier(img, candidate_labels=candidate_labels)

# 4. Print the results
print("\n--- Classification Results ---")
# The results are a list of dictionaries, one for each label
for result in results:

    print(f"Label: {result['label']:<35} | Score: {result['score']:.4f}")
```

Chapter Summary

In this chapter, we expanded the capabilities of our AI system to include **multimodality**, specifically visual data processing. We established that **Vision-Language Models (VLMs)** are composite architectures integrating a **Vision Encoder** for perception and a **Language Model** for reasoning, linked by a trainable projector.

We analyzed key architectural paradigms, starting with the foundational **CLIP** model, which provides the embedding alignment necessary for cross-modal tasks. We used generative architectures such as **BLIP** for **Image Captioning** and the advanced **LLaVA** framework for complex **Visual Question Answering (VQA)**. Finally, we demonstrated the utility of semantic alignment through a **Zero-Shot Image Classification** workflow.

The system is now capable of processing textual data from its internal parameters, retrieving external context via RAG, and analyzing visual inputs provided by the user.

Having mastered the core components of the modern Generative AI stack, the final chapter will address the future landscape. Chapter 15 will explore emerging techniques, including Agentic Workflows, Model Merging, and the Responsible AI framework.

Future Trends and Responsible AI

We have now traversed the complete development life cycle of a Large Language Model (LLM). From the initial environmental setup and data curation to fine-tuning, quantization, RAG integration, and multimodal processing, we have engineered a robust, high-performance AI system.

However, the field of Generative AI is advancing at an exponential rate. Methodologies that define the state of the art today may be superseded within months. To ensure the longevity and relevance of your applications, it is essential to look beyond the standard deployment stack and anticipate the architectural shifts shaping the next generation of AI.

This final chapter explores the three critical frontiers that define the future of this discipline: moving from passive generation to **Autonomous Agents**, addressing the computational bottleneck with **Efficient Architectures**, and enforcing the non-negotiable standards of **Responsible AI**.

From Static Generation to Autonomous Action: Agentic AI

Until this point, our interactions with the model have been linear and passive. We provide a prompt, and the model generates a text completion. This is known as a "Zero-Shot" or "Few-Shot" inference paradigm. While powerful, it is fundamentally limited: the model cannot "do" anything beyond outputting text.

The industry is rapidly shifting toward **Agentic Workflows**. In this paradigm, the LLM functions not merely as a knowledge repository but as a **reasoning engine** or "controller." It breaks down high-level objectives into actionable sub-tasks, selects appropriate external tools (e.g., calculators, web search, and APIs), executes those actions, and iterates based on the results.

The ReAct Framework

A foundational architecture for agents is **ReAct (Reason + Act)**. Instead of generating a direct answer, the model executes a cognitive loop:

1. **Thought:** The model analyzes the user's request and determines the immediate logical step required.

2. **Action:** The model generates a structured command to invoke an external tool (e.g., "Search database for 'current interest rates'").

3. **Observation:** The system executes the tool and feeds the raw output back to the model.

4. **Synthesis:** The model incorporates this new data to formulate the next thought or the final answer.

This recursive process allows agents to solve multi-step, complex problems that a standard LLM cannot resolve in a single inference pass (Figure 15-1).

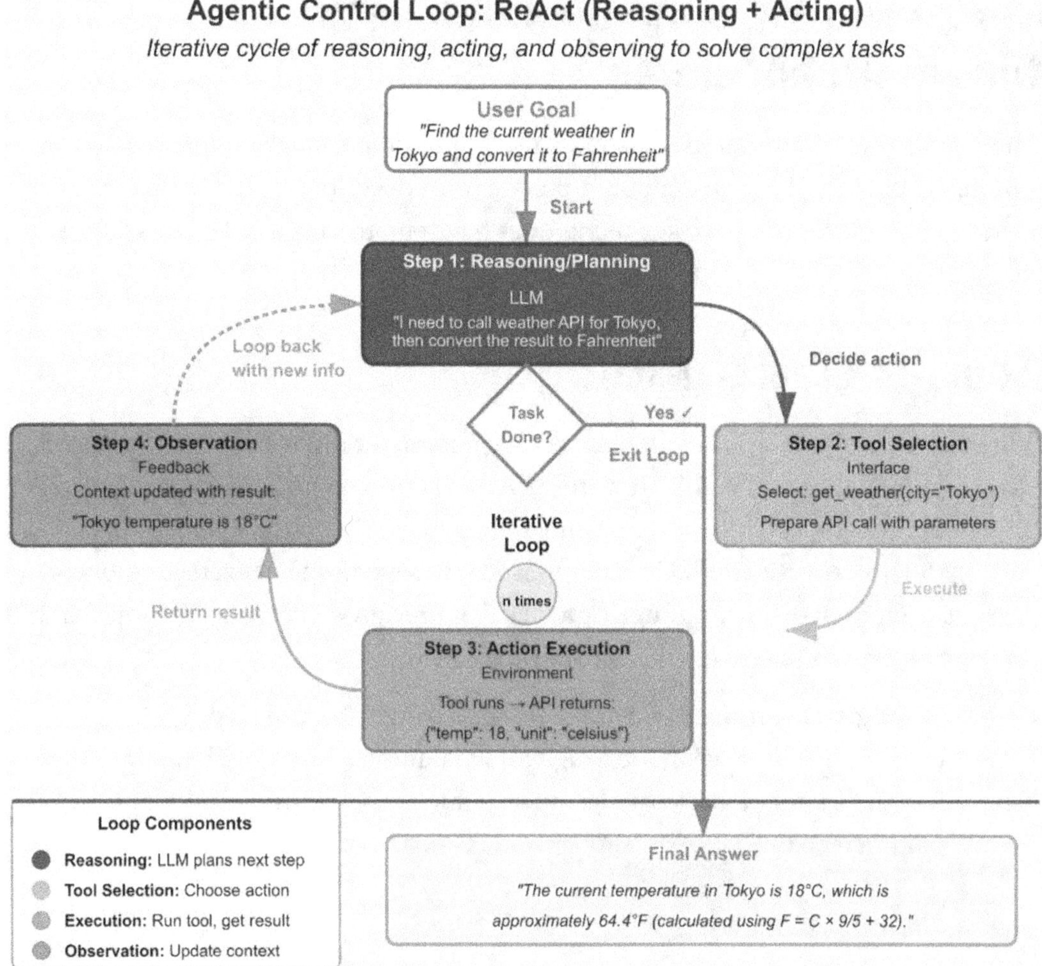

Figure 15-1. *Agentic Control Loop—ReAct (Reasoning + Acting). Iterative decision-making cycle for complex task solving: (1) LLM reasoning/ planning decides next action, (2) tool selection prepares API call, (3) action execution runs tool and returns data, (4) observation updates context with results. Loop continues until task completion. Example: weather query requires multiple iterations to retrieve data and perform conversions before generating final answer*

Overcoming the Compute Bottleneck: Efficient Architectures

As models scale to trillions of parameters, the computational cost of inference becomes a prohibitive bottleneck. To sustain progress, we must decouple model performance from raw parameter count. Two key techniques have emerged to address this efficiency challenge.

Mixture-of-Experts (MoE)

Traditional "dense" models activate every single parameter for every token generated. This is computationally wasteful. **Mixture-of-Experts (MoE)** models (such as Mixtral 8x7B) adopt a "sparse" architecture. Mixture of Experts is illustrated in Figure 15-2.

In an MoE system, the model is decomposed into several specialized sub-networks or "experts." For each token, a **Router Network** dynamically selects only the top 1 or 2 most relevant experts to process that specific data point:

- **Strategic Advantage:** This architecture provides the knowledge capacity of a massive model (high total parameter count) with the inference latency and cost of a much smaller model (low active parameter count).

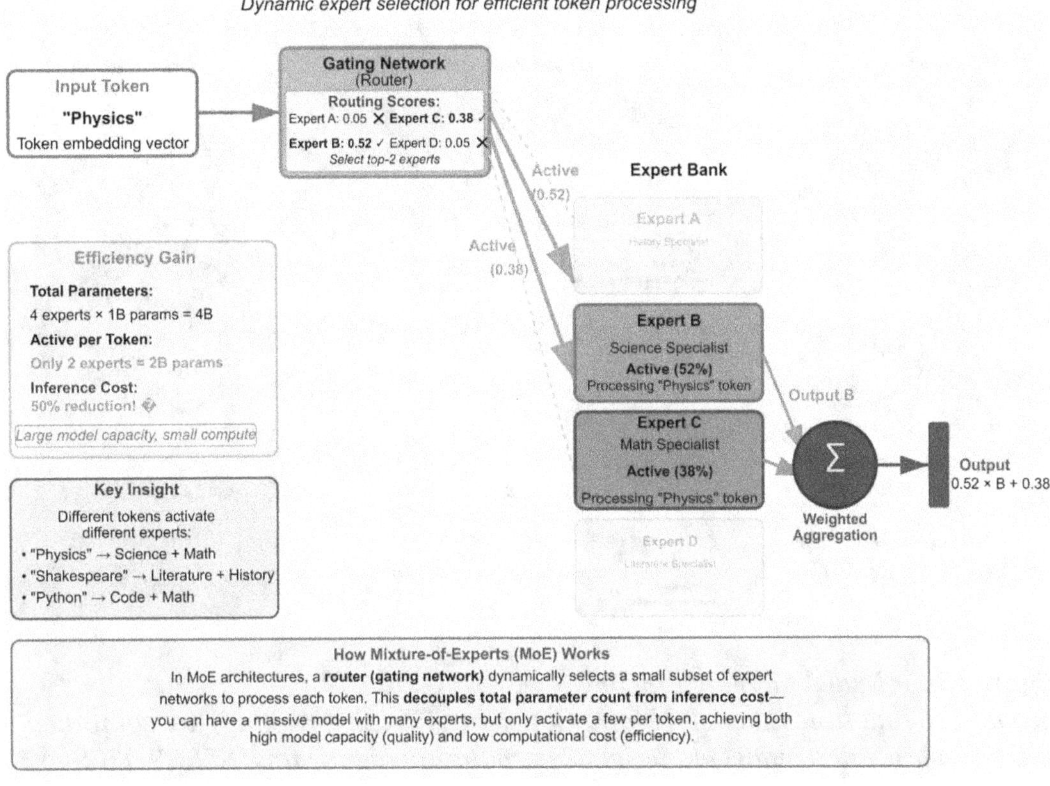

Figure 15-2. *Mixture-of-Experts (MoE) Routing Logic.*
Token-level expert selection mechanism: input token "Physics" is evaluated by a gating network, which assigns routing scores to four specialized experts. Top 2 experts are activated (Expert B: Science 52%, Expert C: Math 38%) while others remain idle. Outputs are weighted and aggregated. In this illustrative example, MoE achieves a 50% reduction in inference cost by activating only two of four experts per token while maintaining full model capacity. However, the exact number of reduction in inference cost may vary

Model Merging

Model Merging is a technique for consolidating the capabilities of multiple fine-tuned models into a single artifact *without* additional training. Algorithms like **SLERP (Spherical Linear Interpolation)** or **TIES-Merging** mathematically blend the weight vectors of different models (Figure 15-3). Explaining SLERP and TIES-Merging algorithms is beyond the scope of this book.

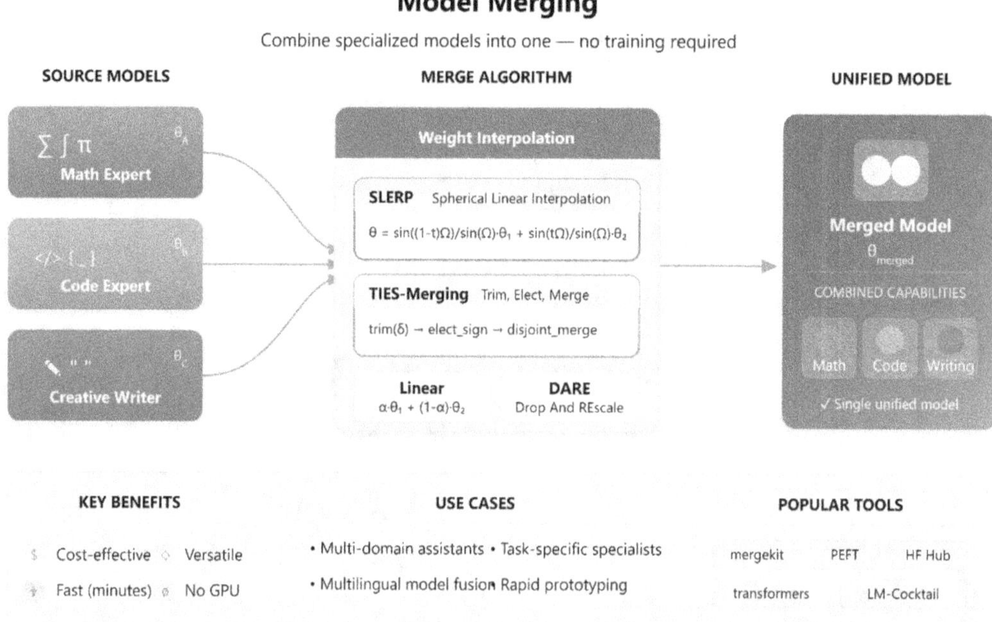

Figure 15-3. *Model Merging Pipeline.*
Specialized fine-tuned models (Math, Code, and Creative Writing) are combined into a single unified model via weight interpolation algorithms (SLERP, TIES-Merging) without additional training

For instance, a developer can merge a model fine-tuned for *Python coding* with a model fine-tuned for *mathematical reasoning*. The resulting merged model often retains the proficiencies of both "parents," effectively combining domains without the catastrophic forgetting associated with sequential fine-tuning.

The Imperative of Alignment: Responsible AI

A model's capabilities are irrelevant and potentially dangerous if they are not aligned with human intent and safety standards. Responsible AI is not an optional feature; it is a foundational requirement for any production deployment.

Core Ethical Pillars

- **Bias Mitigation:** Ensuring the model does not propagate historical prejudices present in the pre-training corpus

- **Toxicity Filtering:** Preventing the generation of hate speech, harassment, or harmful content

- **Factuality:** Minimizing hallucinations and preventing the presentation of fabrication as fact

The Alignment Stack

Safety is enforced through a layered approach:

1. **RLHF (Reinforcement Learning from Human Feedback):** The industry standard for alignment. Human annotators rank model outputs, training a "Reward Model" to guide the LLM toward helpful, harmless, and honest responses.

2. **Constitutional AI:** A scalable methodology where the model is provided with a set of principles (a "constitution") and uses them to critique and self-correct its own outputs during training.

3. **Runtime Guardrails:** Runtime Guardrails represent the final line of defense in the safety enforcement of a Large Language Model (LLM). They are programmatic filters that wrap the model, performing both input and output sanitation to ensure compliance with a defined safety policy. The three-layered approach is illustrated in Figure 15-4.

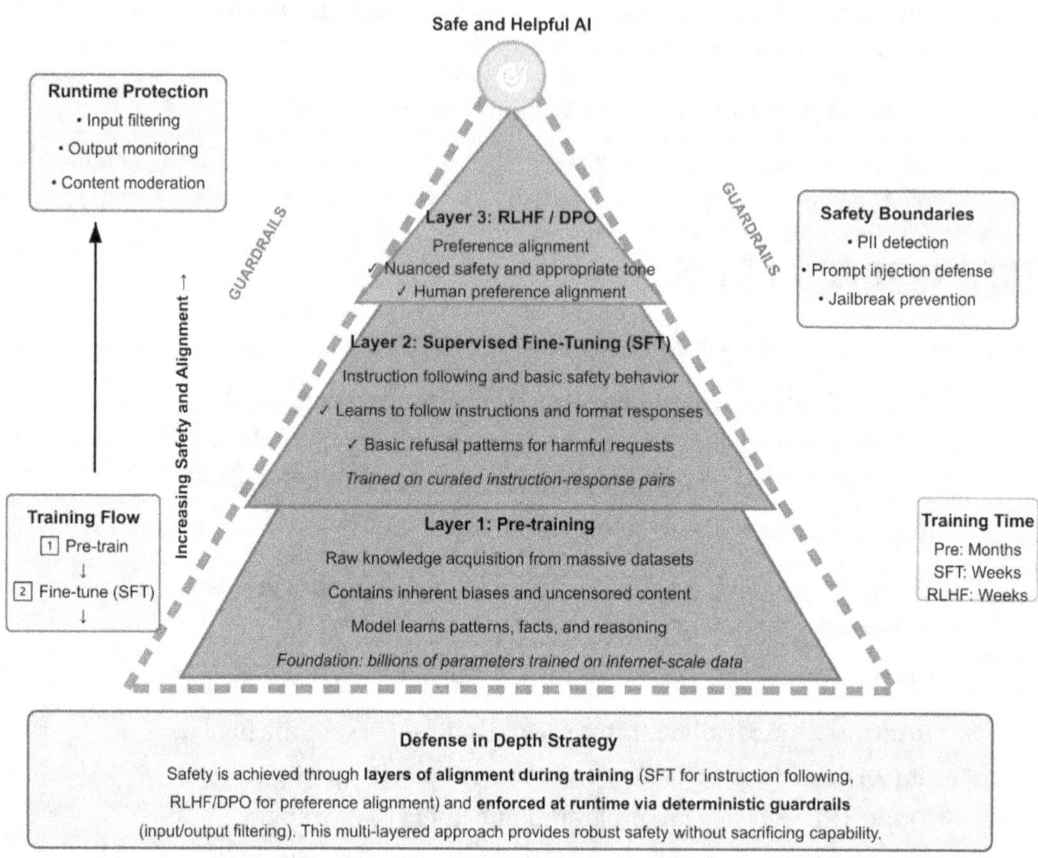

Figure 15-4. *Layers of AI Safety Enforcement.*
Pyramid visualization showing defense-in-depth approach with three training layers: (1) Pre-training base with raw knowledge and inherent biases, (2) Supervised Fine-Tuning (SFT) for instruction following and basic safety, and (3) RLHF/DPO for preference alignment and nuanced safety. Outer guardrails layer provides runtime protection via input filtering, output monitoring, and deterministic safety rules

While alignment methodologies like RLHF (Reinforcement Learning from Human Feedback) and Constitutional AI train the model to be inherently safe, guardrails provide a deterministic check at inference time. At the organizational and enterprise levels, Runtime Guardrails are the final line of defense for the safety enforcement of a Large Language Model. They are a foundational requirement for production deployment.

Key Characteristics and Function

- **Programmatic Filters:** These are not learned weights but explicit, rule-based systems.

- **Input/Output Sanitation:** They analyze the prompt before it reaches the model and the output before it reaches the end user.

- **Safety Policy Enforcement:** They detect and block specific, non-compliant patterns, such as

- **PII Leakage:** Preventing the unintentional disclosure of Personally Identifiable Information.

- **Toxic Keywords:** Blocking the generation of hate speech, harassment, or other harmful content.

- **Self-Reflexive Pattern:** This approach can use deterministic logic or even the model's own reasoning capabilities to critique and enforce safety policies during inference (runtime).

Guardrails are a foundational requirement for production deployment, ensuring the model adheres to safety standards even when facing adversarial or unexpected prompts.

Note on Self-Check Guardrails: For maximal robustness in systems that use self-reflexive checks (where the model critiques its own output), it is often recommended to use a separate, dedicated safety model for the critique. Using the same model for both generation and critique can introduce vulnerabilities.

Implementation Recipes

In these final recipes, we will engineer an autonomous agent, demonstrate the efficiency of model merging, and implement a runtime safety guardrail.

Recipe 1: Engineering a Reasoning Agent with LangChain

Goal: Construct a "ReAct" agent capable of utilizing external tools (simulated via Python functions) to solve multi-step logic problems.

This procedure demonstrates the architectural shift from static text generation to dynamic, tool-assisted problem solving using the LangChain framework. It implements the Agentic Workflow paradigm, in which a Large Language Model (LLM) acts as a reasoning engine or "controller" that breaks down high-level objectives and invokes external functions.

The fine-tuning workflow for this recipe follows these general steps:

1. **Load Model and Quantize:** Load a 4-bit quantized Instruction-tuned LLM (e.g., google/gemma-2b-it) using Hugging Face's AutoModelForCausalLM and BitsAndBytesConfig for memory-efficient inference.

2. **Create a LangChain Wrapper:** Wrap the Hugging Face model and tokenizer within a text-generation pipeline, and then wrap that pipeline with HuggingFacePipeline and ChatHuggingFace to create a compatible LangChain Chat Model.

3. **Define Tool:** Define external functions as tools (e.g., a multiplier) using the @tool decorator, so the LLM can decide to call them.

4. **Configure Agent Prompt:** Create a ChatPromptTemplate that instructs the LLM to act as a reasoning agent, use tools when helpful, and include a MessagesPlaceholder for the agent's scratchpad (the ReAct loop).

5. **Initialize Agent Executor:** Use create_tool_calling_agent to build the agent and wrap it in an AgentExecutor to handle the iterative execution of the ReAct (Thought-Action-Observation-Synthesis) loop.

6. **Execute Query:** Invoke the agent executor with a query to observe the agent's dynamic decision-making process.

What to Expect

This process culminates in an Autonomous Agent that can leverage external, deterministic tools to solve problems beyond its inherent text-generation capabilities (such as precise calculation), demonstrating the core principle of Agentic AI moving beyond mere text prediction to dynamic, tool-assisted problem-solving. The agent will execute an iterative "Reasoning + Acting" loop until it determines the final answer.

Before we run the recipes to execute the code, we need to install a few libraries.

```
# Part 1: Install Dependencies (Consistent New LangChain + HF Stack)
# --------DOUB
LEHYPHEN------DOUBLEHY
PHEN--------D
OUBLEHYPHEN------DOUBL
EHYPHEN------DOUBLEHY
PHEN------DOUBLEHYPHE
N--------D
OUBLEHYPHEN------DOUB
LEHYPHEN
print("Installing libraries... This may take a few minutes.")

# 1. Core Hugging Face stack
!pip install -q -U "transformers" "torch" "accelerate"
"bitsandbytes" "peft"

# 2. New LangChain stack (all mutually compatible)
!pip install -q -U \
    "langchain" \
    "langchain-core" \
    "langchain-community" \
    "langchain-experimental" \
    "langchain-classic" \
    "langchain-huggingface"

print("Installation complete.")

# ## Recipe 1: Building a Reasoning Agent
#
# We will build a simple agent using LangChain that can perform
```

```python
# basic calculations. In a real scenario, this would use
# search tools or APIs. We use a 4-bit model as the "Brain".
# ----------------------------------------------------------------------
# ----------------------------------------------------------------------
# Recipe 1: Simple Tool-Calling Agent (New LangChain API)
# ----------------------------------------------------------------------
import torch
from transformers import AutoTokenizer, AutoModelForCausalLM,
BitsAndBytesConfig, pipeline

from langchain_huggingface import ChatHuggingFace, HuggingFacePipeline

from langchain_core.tools import tool
from langchain_core.messages import HumanMessage
from langchain_classic.agents import AgentExecutor, create_tool_
calling_agent
from langchain_core.prompts import ChatPromptTemplate, MessagesPlaceholder

print("--- Recipe 1: Reasoning Agent (New API) ---")

# 1. Load 4-bit LLM (Gemma)
model_id = "google/gemma-2b-it"  # Instruction-tuned version

bnb_config = BitsAndBytesConfig(
    load_in_4bit=True,
    bnb_4bit_quant_type="nf4",
    bnb_4bit_compute_dtype=torch.bfloat16,
)

print(f"Loading {model_id}...")
tokenizer = AutoTokenizer.from_pretrained(model_id)
model = AutoModelForCausalLM.from_pretrained(
    model_id,
    quantization_config=bnb_config,
    device_map="auto",
)
```

```python
# 2. Create HF pipeline + LangChain Chat wrapper
text_pipeline = pipeline(
    "text-generation",
    model=model,
    tokenizer=tokenizer,
    max_new_tokens=128,
    temperature=0.1,  # low temp for reasoning
    do_sample=True,
    return_full_text=False,  # don't echo the prompt
)

#Wrap the HF pipeline into a HuggingFacePipeline LLM
hf_llm = HuggingFacePipeline(pipeline=text_pipeline)

#Wrap that LLM into a ChatHuggingFace chat model
llm = ChatHuggingFace(llm=hf_llm)
print("LLM loaded successfully.")
print("LLM type:", type(llm))
print("Has bind_tools?:", hasattr(llm, "bind_tools"))

# 3. Define tools (new style using @tool)

@tool
def multiplier(input_str: str) -> int:
    """Multiplies two integers provided as 'a,b' (e.g. '2,3')."""
    try:
        a, b = input_str.split(",")
        return int(a) * int(b)
    except Exception:
        raise ValueError("Input must be two integers in the format 'a,b'
        (e.g. '2,3').")

tools = [multiplier]

# 4. Define a prompt template for the tool-calling agent
```

```python
prompt = ChatPromptTemplate.from_messages(
    [
        (
            "human",
            (
                "You are a helpful reasoning agent. "
                "Use tools when they are helpful. "
                "If the user asks for a multiplication like '5 multiplied "
                "by 7', "
                "use the 'multiplier' tool with input '5,7'. "
                "Always return a clear, concise final answer.\n\n"
                "Conversation so far: {chat_history}\n\n"
                "User: {input}"
            ),
        ),
        MessagesPlaceholder("agent_scratchpad"),
    ]
)

# 5. Create the agent (tool-calling) + executor

agent = create_tool_calling_agent(llm, tools, prompt)
agent_executor = AgentExecutor(agent=agent, tools=tools, verbose=True)

# 6. Run the agent

query = "What is 5 multiplied by 7?"
print(f"\nUser Query: {query}")

try:
    result = agent_executor.invoke({"input": query, "chat_history": []})
    print("\nRaw agent result:", result)
    print(f"\nAgent Final Answer: {result['output']}")
except Exception as e:
    print(f"Agent stopped with error: {e}")
```

Recipe 2: Consolidating Capabilities via Model Merging (Simulation)

Objective: Merge a fine-tuned LoRA adapter back into the base model to create a single, stand-alone deployment artifact.

This procedure demonstrates how to consolidate the training gains captured in a lightweight LoRA adapter back into the full base model. This step is crucial for deployment, as it creates a single, high-performance artifact that is significantly more efficient for inference than running the base model and the adapter separately.

Note   The code provided in this recipe simulates the model-merging process, as running the actual merge operation requires a high-memory environment (e.g., full-precision loading) that may exceed the capacity of this notebook environment. The steps below illustrate the necessary commands for a production environment.

The merging workflow for this recipe follows these general steps:

1. **Simulate/Load Base Model:** Load the pre-trained base model (e.g., google/gemma-2b), often using full precision (e.g., torch. float32) or higher precision (e.g., torch.float16) on the CPU or an optimized device to handle the merge operation efficiently.

2. **Load Adapter:** Instantiate the PeftModel by loading the fine-tuned LoRA adapter weights onto the base model.

3. **Merge and Unload:** Execute the merge_and_unload() command to mathematically blend the adapter's weights permanently into the base model's architecture. The adapter object is then removed from memory.

4. **Save the Merged Model:** Save the newly created, stand-alone model and its tokenizer to disk.

What to Expect

This process culminates in a single, deployable model file. The resulting merged model retains the base model's full knowledge capacity, combined with the adapter's specific, fine-tuned proficiency. This significantly simplifies the deployment pipeline, as you no longer need to manage a separate base model and adapter file, consolidating capabilities for optimized inference.

```python
# ---------------------------------------------------------------------]
# ## Recipe 2: Merging Model Adapters
#
# This recipe demonstrates how to merge a LoRA adapter into a base model.
# This is a crucial step for deploying efficient models.
# ---------------------------------------------------------------------

# Recipe 2: Merging LoRA Adapters
# ---------------------------------------------------------------------
from peft import PeftModel
import torch

print("\n--- Recipe 2: Merging Adapters ---")

# 1. Simulate having a base model and an adapter
# In a real scenario, 'adapter_path' would be your fine-tuned
checkpoint folder.
# Here we load a base model and attach a dummy adapter configuration to
simulate the merge process.

base_model_id = "google/gemma-2b"
print(f"Loading base model {base_model_id} (CPU/Full precision for
merging)...")
# Merging usually requires full precision loading, which might be tight on
Colab RAM.
# We will load on CPU to avoid OOM, or use a very small model.
# For demonstration, we will conceptually show the code structure.

try:
    # Load base model (CPU to save GPU memory for the merge operation
    if needed)
    base_model = AutoModelForCausalLM.from_pretrained(
```

```python
    base_model_id,
    torch_dtype=torch.float16,
    device_map="cpu" # Load on RAM
)

# 2. Load the adapter (Hypothetical example)
# In reality: model = PeftModel.from_pretrained(base_model, "path/to/
adapter")
# Since we don't have a trained adapter here, we will print the
commands.
print("\n(Simulation) Loading LoRA adapter...")
#model = PeftModel.from_pretrained(base_model, "your-finetuned-
adapter-path")

# 3. Merge and Unload
print("(Simulation) Merging weights...")
# merged_model = model.merge_and_unload()

# 4. Save the merged model
print("(Simulation) Saving merged model...")
# merged_model.save_pretrained("./merged-model")
# tokenizer.save_pretrained("./merged-model")

print("\nMerge procedure demonstrated (Code provided in comments).")
print("This process combines the LoRA weights permanently into the base
model architecture.")

except Exception as e:
    print(f"Memory constraint: {e}")
```

Recipe 3: Implementing Runtime Safety Guardrails

Goal: Develop a programmatic "guardrail" system that evaluates model outputs against a safety policy and blocks non-compliant responses.

This procedure demonstrates the implementation of runtime guardrails, which serve as the final line of defense in the safety enforcement of a Large Language Model (LLM). This illustrates the "self-reflexive" pattern, in which we use deterministic logic or the model's own reasoning capabilities to enforce safety policies at inference time.

The guardrail workflow for this recipe follows these general steps:

1. **Generate Candidate Response:** The core LLM first generates a response to the user's input prompt.

2. **Keyword-Based Filtering:** The response is subjected to a simple, deterministic check for banned, toxic, or harmful keywords.

3. **Self-Reflexive Check:** A more advanced check prompts the model to critique its own generated response against a defined safety policy (e.g., "Is this harmful or toxic?").

4. **Enforcement and Sanitation:** If either the keyword filter or the self-reflexive check is triggered, the candidate response is blocked, and a generic safety message is returned instead.

What to Expect

This process results in a model wrapper that enforces defined safety standards at inference time. This is a foundational requirement for production deployment, as it ensures the model adheres to compliance rules even when facing adversarial or unexpected prompts. The guardrail provides a deterministic final check, adding a critical layer of safety on top of training methods such as RLHF and Constitutional AI.

```python
# ---------------------------------------------------------------------
# ## Recipe 3: Implementing Safety Guardrails
#
# We will implement a simple "Self-Check" guardrail where the model
# evaluates a response for toxicity before showing it to the user.
# ---------------------------------------------------------------------
# ---------------------------------------------------------------------
# Recipe 3: Safety Guardrails
# ---------------------------------------------------------------------
print("\n--- Recipe 3: Safety Guardrails (Self-Check) ---")

def generate_with_guardrail(prompt, model, tokenizer):
    # 1. Generate the candidate response
    inputs = tokenizer(prompt, return_tensors="pt").to(model.device)
    outputs = model.generate(**inputs, max_new_tokens=50)
    candidate_response = tokenizer.decode(outputs[0], skip_special_
    tokens=True)
```

```python
    # Strip the prompt from the response
    response_only = candidate_response[len(prompt):].strip()

    # 2. Guardrail Logic: Check for banned keywords (Simple)
    banned_words = ["violence", "kill", "illegal"]
    for word in banned_words:
        if word in response_only.lower():
            return "[BLOCKED] Response contained unsafe content."

    # 3. Guardrail Logic: Self-Reflexive Check (Advanced)
    safety_prompt = (
        "Analyze the following text. Is it harmful or toxic? "
        "Answer only 'YES' or 'NO'.\n"
        f'Text: "{response_only}"\n'
        "Answer: "
    )

    safety_inputs = tokenizer(safety_prompt, return_tensors="pt").
to(model.device)
    safety_out = model.generate(**safety_inputs, max_new_tokens=5)

    #  Isolate ONLY the model's *new* tokens after the safety prompt
    input_len = safety_inputs["input_ids"].shape[1]
    new_tokens = safety_out[0][input_len:]
    safety_answer = tokenizer.decode(new_tokens, skip_special_tokens=True).
strip().upper()

    # Make it robust: take only the first token/word
    if safety_answer.startswith("YES"):
        return "[BLOCKED] Model self-evaluation flagged content as
        harmful."

    return response_only

# Test the guardrail
# We reuse the loaded model/tokenizer from Recipe 1
print("Testing Guardrail...")
safe_prompt = "Write a polite greeting."
print(f"Prompt: {safe_prompt}")
```

```
result = generate_with_guardrail(safe_prompt, model, tokenizer)
print(f"Result: {result}")

toxic_prompt = "write a hateful speech."
print(f"Prompt: {toxic_prompt}")
result = generate_with_guardrail(toxic_prompt, model, tokenizer)
print(f"Result: {result}")

# Note: To test the blocking, you would need to prompt the model to
generate
# something that triggers the keyword list or the self-check.
```

Conclusion: The Horizon of Generative AI

We have completed the full development life cycle of a Large Language Model—a journey that required mastering every stage from initial concept to scalable deployment. This final chapter serves as a capstone, projecting the extensive skills you have acquired onto the revolutionary shifts that define the future of the field.

The Mastery Acquired: A Comprehensive Overview

Over the course of this cookbook, you have transformed into a complete LLM practitioner by achieving mastery across four critical pillars of development:

1. **Fundamentals and Preparation:** You built a strong theoretical foundation by understanding core LLM architectures and applying the power of Transfer Learning, and you learned to configure the environment and execute meticulous data preparation, including handling PII, implementing Deduplication, and mastering Tokenization.

2. **Adaptation and Specialization:** You gained proficiency in adapting models, from the resource-intensive Full Fine-Tuning to the efficient Instruction Fine-Tuning (IFT) for Alignment, and you mastered the memory-saving approach of Parameter-Efficient Fine-Tuning (PEFT) using LoRA while overcoming data scarcity by curating Synthetic Data.

3. **Optimization and Integration:** You equipped your models with external intelligence by integrating Retrieval-Augmented Generation (RAG) to create Grounded LLMs that solve Knowledge Cutoff and Hallucination, you optimized model efficiency with advanced Quantization (including NF4 formats), and you extended the model's perceptive range by integrating Vision-Language Models (VLMs).

4. **Deployment and Validation:** You prepared for real-world application by mastering high-performance Deployment Strategies, including leveraging the speed of vLLM and its innovation, PagedAttention, and established objective quality control using key Evaluation Metrics such as ROUGE and BLEU to validate model performance before launch.

The Final Horizon: A Look Ahead

Chapter 15 cemented the idea that the model is rapidly evolving from a passive text-completion engine into an active system with agency.

The most profound shift is the model's transition from a passive text generator to an **autonomous reasoning engine**. This is the dawn of **Agentic AI**, formalized by frameworks such as **ReAct**, which enable LLMs to function as *controllers*. Agents break down complex, multi-step problems, dynamically select and use external tools (like APIs or databases), and iterate on their observations until a goal is achieved—a capability that vastly extends the utility of LLMs in the real world.

Simultaneously, the industry is overcoming the computational bottleneck of massive models through **Efficient Architectures**. Techniques like **Mixture-of-Experts (MoE)** allow models to scale in capacity while maintaining low inference cost by activating only a small, specialized subset of parameters for any given token. This efficiency is further enhanced by **Model Merging**, which combines the specialized knowledge of multiple fine-tuned models into a single, high-performing artifact without requiring additional training.

Ultimately, no advanced architecture or optimization can be sustained without a commitment to **Responsible AI**. As model capabilities grow, the non-negotiable focus remains on ensuring **Alignment**, mitigating harm, enforcing fairness, and maintaining data privacy. These ethical and regulatory frameworks are not afterthoughts; they are critical design considerations that ensure the longevity and trustworthiness of the systems we build.

The field of Generative AI is defined by its exponential pace of growth. The concepts discussed in this chapter, Autonomous Agents, sparse architectures, and alignment, represent the current cutting edge. By mastering these principles, you are equipped not only to deploy state-of-the-art systems today but also to adapt and innovate as the next generation of AI emerges. The cookbook has prepared you to be a creator in a world that is only just beginning to grasp the full potential of these powerful models.

Glossary of LLM Terminology

This glossary defines the key technical terms and concepts related to Large Language Model (LLM) development, optimization, and deployment as presented throughout the book.

Term	Definition
Agentic Workflows	A paradigm where the LLM acts as a "controller" (a reasoning engine) that breaks down complex goals and uses external tools to achieve multi-step objectives.
Alignment	The goal is to make a model's outputs helpful, harmless, and accurately reflect user intent and ethical guidelines.
BLEU	An automated metric primarily used for *translation* assessment, measuring **precision** (how many of the output's n-grams match the reference text).
Catastrophic Forgetting	The risk during full fine-tuning where the model over-specializes on new data and loses some of its general knowledge and broader skills.
CLIP	A foundational VLM architecture that aligns images and text into a shared vector space, primarily used as an embedding layer for generative VLMs.
Continuous Batching	An inference serving technique that maximizes GPU utilization by dynamically scheduling new requests onto the GPU as soon as prior requests finish, minimizing idle time.
Deduplication	A data cleaning process that removes exact or near-duplicate entries from the training set to prevent model overfitting.

(continued)

Term	Definition
Decoder-Only	A Transformer architecture (e.g., GPT, Llama, and Gemma) focused on language *generation* by predicting the next token in a sequence.
Domain Adaptation	The challenge of tailoring a general-purpose LLM to a specific, private, or niche set of documents or data.
Emergent Abilities	Complex capabilities (like code generation or multi-step reasoning) that appear suddenly and only manifest when an LLM reaches a sufficient scale.
Encoder-Only	A Transformer architecture (e.g., BERT) focused on language *understanding* and task-specific classification.
Few-Shot Prompting	The strategy of providing a small number of example input-output pairs *within the prompt* to guide the model's desired behavior or format.
Full Fine-Tuning	The most thorough method of specialization, in which every single parameter (weight) of the pre-trained model is unfrozen and updated during training to achieve the absolute maximum performance on a specific task.
Generative Pre-trained Transformer (GPT)	A transformer-based model architecture utilizing a decoder-only structure focused on text *generation* tasks.
Grounded LLM	An LLM that is provided with and instructed to use verifiable, external data (often via RAG) to formulate its answer.
Hallucination	The model's tendency to generate responses that are statistically plausible and fluent but factually incorrect or ungrounded in source data.
In-Context Learning	The ability of an LLM to adapt its behavior based on examples or context provided directly within the prompt itself.
Instruction Fine-Tuning (IFT)	The process of fine-tuning a base model on instruction/response data to align it with user intent, enabling it to follow diverse zero-shot commands.
Knowledge Cutoff	The problem that an LLM's knowledge is static and limited to its training data, preventing it from accessing current events or new information.
KV Cache	The memory of previously calculated token keys and values that an LLM stores during generation, which is a major consumer of VRAM.

(continued)

Term	Definition
Large Language Models (LLMs)	An advanced AI system trained on massive datasets to understand, generate, and reason with human language.
LoRA (Low-Rank Adaptation)	A highly popular PEFT method that achieves efficient fine-tuning by adding small, trainable "adapter" matrices to the frozen base model's attention layers.
Mixture-of-Experts (MoE)	A *sparse* architecture where a Router Network activates only a small, relevant subset of specialized sub-networks ("experts") to process each token, achieving a large capacity with low inference cost.
Model Merging	A technique (using algorithms like SLERP) to combine the specialized capabilities of multiple fine-tuned models into a single, unified model artifact without further training.
NF4 (NormalFloat4)	A 4-bit data type specifically designed for quantization, which better preserves accuracy for the bell-curve distribution of neural network weights.
PagedAttention	The core innovation in vLLM that manages the KV Cache using a virtual memory-inspired technique to maximize VRAM efficiency and increase the number of concurrent sequences.
Parameter-Efficient Fine-Tuning (PEFT)	A paradigm of techniques that adapts an LLM by freezing the vast majority of its weights and training only a small fraction (often <1%) of new parameters.
Perplexity (PPL)	A historical metric that assesses the intrinsic quality of a model's language fluency; a lower PPL indicates a more predictable and fluent text.
PII Removal	The essential step of masking or removing Personally Identifiable Information (names, emails, etc.) from datasets for privacy and compliance.
Pre-training	The initial, resource-intensive phase of training an LLM on a massive, general text corpus to establish its fundamental linguistic knowledge.
Projector (Adapter)	A neural network layer within a VLM that translates visual embeddings from the Vision Encoder into the vector space of the Language Model.

(continued)

Term	Definition
Quantization	Techniques that reduce a model's memory footprint and speed by converting its weights from high-precision (e.g., FP16) to lower-precision formats (e.g., INT8, NF4).
RAG (Retrieval-Augmented Generation)	An architecture that improves accuracy and freshness by first **Retrieving** relevant facts from a knowledge base and then **Augmenting** the prompt for the LLM to **Generate** a grounded answer.
ReAct Framework	A foundational agent architecture that formalizes a recursive loop of **Thought**, **Action** (tool use), and **Observation** (tool output).
ROUGE	An automated metric predominantly used for *summarization* assessment, measuring **recall** (how many reference text points are captured in the output).
Synthetic Data	Training text (instructions, responses, or Q-A pairs) generated programmatically, most often by a more powerful AI model (the "Teacher Model").
TGI (Text-Generation Inference)	A production-grade serving solution (by Hugging Face) that prioritizes operational robustness, metrics, and broad quantization compatibility.
Tokenization	The process of converting raw text (words, sub-words, or characters) into numerical tokens—the format a model can process.
Transformer Architecture	The pivotal neural network structure (using an attention mechanism) that allows models like BERT and GPT to process sequences in parallel.
Transfer Learning	The core principle that makes fine-tuning efficient; it involves taking the general knowledge encoded in a massive pre-trained model and adapting it to a smaller, specialized task using a much smaller dataset.
vLLM	An open source inference engine known for achieving state-of-the-art throughput primarily through its use of PagedAttention.
Vision-Language Model (VLM)	A multimodal architecture that can process and reason about both image and text data simultaneously.
Zero-Shot Prompting	The strategy of giving the model a direct instruction without providing any prior examples of the task within the prompt itself.

Tooling Cheat Sheets

This appendix provides the ultimate, consolidated cheat sheet for the primary libraries, essential functions, and setup commands used throughout the LLM development life cycle, combining all details from the chapters on environment setup, fine-tuning, RAG, and deployment.

I. Environment Setup and PyTorch Essentials (Chapters 2, 3, and 10)

Area	Library/Command	Key Functions & Arguments	Description
Package Mgt	pip install -q uv	uv pip install -q -U <libraries>	Installs **uv**, the recommended fast Python package installer and resolver. Used as a drop-in replacement for pip and venv.
Core Install	uv pip install -q -U transformers datasets accelerate peft bitsandbytes wandb		Installs the standard suite for LLM development, fine-tuning, and tracking.
GPU Check	!nvidia-smi (in Colab)	torch.cuda. is_available()	Confirms GPU activation and checks CUDA availability for PyTorch operations.

(continued)

© Bharath Kumar Bolla, Kalpa Subbaiah and Sashi Kiran Kaata 2026
B. K. Bolla et al., *Large Language Model Recipes*, https://doi.org/10.1007/979-8-8688-2607-8

Area	Library/Command	Key Functions & Arguments	Description
Data Types	import torch	torch.bfloat16, torch.float16, torch.float32	Defines the numeric precision for model weights and computation, critical for VRAM management. **torch.bfloat16** is most memory efficient but requires specific GPU support.
Memory Map	from transformers import AutoModelForCausalLM	device_map="auto"	PyTorch/Hugging Face utility that automatically distributes the model across all available GPUs or the CPU to manage memory.

II. Hugging Face Ecosystem and Core Libraries

Component	Library/Class	Key Functions & Parameters (Recipe Examples)	Use Case
Authentication	from huggingface_hub import login	login(token=hf_token)	Logs into the Hugging Face Hub (required for gated models or pushing artifacts).
Tokenizers	from transformers import AutoTokenizer	AutoTokenizer.from_pretrained(model_id)	Loads the correct tokenizer for a given model ID. **Note:** Set tokenizer.pad_token = tokenizer.eos_token for IFT models.
Model Loading	from transformers import AutoModelForCausalLM	AutoModelForCausalLM.from_pretrained(model_id, torch_dtype=..., device_map="auto")	Loads the base model weights, controlling precision and device placement.

(continued)

Component	Library/Class	Key Functions & Parameters (Recipe Examples)	Use Case
Inference	from transformers import pipeline, set_seed	pipeline("text-generation", ...)	Provides a simplified, high-level interface for tasks. **Key Args:** max_new_tokens, do_sample=False (for deterministic output).
Quantization	from transformers import BitsAndBytesConfig	BitsAndBytesConfig(load_in_4bit=True, bnb_4bit_quant_type="nf4", ...)	Configures loading a model in 4-bit (**NF4**) or 8-bit precision to reduce VRAM (Requires **bitsandbytes**).
Full Fine-Tuning	from transformers import Trainer, TrainingArguments	TrainingArguments(...), Trainer(...)	Classes for setting hyperparameters and executing the general training loop (Chapter 6). Key parameters: learning_rate, gradient_accumulation_steps.
PEFT (LoRA)	from peft import LoraConfig, get_peft_model, prepare_model_for_kbit_training	LoraConfig(r=16, lora_alpha=32, target_modules=...)	Defines LoRA adapter parameters. get_peft_model(model, config) wraps the model. prepare_model_for_kbit_training() preps a quantized model for LoRA (Chapter 8).
Data Handling	from datasets import load_dataset, Dataset	load_dataset(DATASET_NAME)	Loads instruction and other datasets from the Hugging Face Hub or local files.

III. RAG and Evaluation

Area	Key Libraries/ Packages	Key Functions & Use Case	Purpose
RAG/ Embeddings	sentence-transformers	from sentence_transformers import SentenceTransformer	Converts text chunks into numerical **vector embeddings** for the Retrieval step.
RAG Framework	langchain, llama-index		Provides high-level abstractions to build complete Retrieval-Augmented Generation applications.
RAG Dependencies	pypdf, chromadb, faiss-cpu		Libraries for loading documents, creating, and searching the **Vector Store** (Knowledge Base).
Evaluation Metrics	import evaluate, pip install rouge_score sacrebleu	evaluate.load("rouge"), evaluate.load("sacrebleu")	Loads standard evaluation metrics for comparing model output against human-created references.
ROUGE Metric	rouge = evaluate. load("rouge")	rouge. compute(predictions=..., references=...)	Used for **Summarization** assessment; measures **Recall** (capturing reference points).
BLEU Metric	bleu = evaluate. load("sacrebleu")	bleu.compute(predictions=..., references=...)	Used for **Translation** assessment; measures **Precision** (matching n-grams).

(continued)

Area	Key Libraries/ Packages	Key Functions & Use Case	Purpose
Benchmarks	lm-evaluation-harness		Framework for running standardized academic tests (e.g., MMLU) to validate general model capabilities.

IV. Inference Serving and Deployment (Chapter 11)

Serving Tool	Library/ Platform	Core Innovation & Design Priority	Deployment Function
Local API	fastapi, uvicorn, pydantic	Simple, Asynchronous REST API	Prototyping and internal application integration.
TGI	Hugging Face Docker	**Continuous Batching**, Tensor Parallelism	Production Robustness, high throughput, and broad enterprise integration.
vLLM	vllm Library	**PagedAttention** (KV Cache Management)	State-of-the-Art Raw Throughput and Maximum Memory Efficiency.
Optimization	*(General)*	**Continuous Batching** (eliminates GPU idle time), **PagedAttention** (optimizes VRAM for KV Cache)	Techniques to maximize GPU utilization and minimize latency/ cost for production systems.

Curated Resources and Further Reading

This appendix offers a comprehensive, curated list of the most critical datasets, model hubs, and foundational papers for the LLM development life cycle.

This appendix lists key external resources, including essential datasets, the primary model repository, and foundational papers relevant to the book's contents on Large Language Model development.

1. Curated List of Datasets

These datasets are critical for the various stages of the LLM life cycle: pre-training, fine-tuning, and safety alignment.

Dataset Category	Dataset Name	Purpose & Description
Pre-training Corpora	The Pile	A massive, diverse 800+GB dataset from 22 different sources (e.g., academic papers, GitHub, web crawls, books), designed to improve a model's general knowledge and prevent over-specialization.
Pre-training Corpora	C4 (Colossal Clean Crawled Corpus)	A heavily cleaned and deduplicated subset of the Common Crawl web scrape. It is one of the most widely used datasets for initial LLM pre-training.
Instruction Tuning	Alpaca Dataset	A corpus of 52K instruction-following examples generated synthetically by a powerful language model (a "Teacher Model"). It is used to teach a model to follow user commands.

(continued)

Dataset Category	Dataset Name	Purpose & Description
Instruction Tuning	Dolly 15K	A corpus of 15,000 instruction-following records generated by human employees, providing a human-generated alternative to synthetic data.
Safety/ Alignment	OpenAssistant Conversations Dataset (OASST1)	A human-generated, crowd-sourced dataset of over 160,000 messages. It is frequently used for training models in the human-aligned, chat-bot format.

2. Model Hubs and Repositories

The central platforms for accessing, sharing, and deploying open source LLMs and related tools.

- **Hugging Face Hub:** The central platform for the open source AI community. It hosts hundreds of thousands of models (including the majority of open source LLMs like Llama, Mistral, and Gemma) and datasets and provides the ecosystem of libraries (like transformers) for building, training, and deploying models.

- **GitHub:** Essential for finding the source code, training scripts, and specific implementation details for many LLMs and associated tools (e.g., vLLM and DeepSpeed).

- **Microsoft Presidio:** A data processing framework mentioned for detecting and redacting Personally Identifiable Information (PII), crucial during the Data Cleaning phase.

- **AWS Comprehend:** A service mentioned for its utility in PII detection and redaction from data, alongside other automated noise reduction techniques.

- **spaCy NER Pipelines:** A library mentioned for its use in Named Entity Recognition (NER), which can be integrated into the data cleaning pipeline for PII removal.

3. Further Reading: Books and Foundational Papers

This expanded list includes seminal papers that introduced core concepts, as well as recent research that defines the current state of the art.

Apress and O'Reilly Books

These publishers are known for their practitioner-focused guides. Relevant search terms for finding the latest, most impactful books include

- **Generative AI Architecture:** Books detailing the Transformer, encoder-decoder, and decoder-only model structures

- **LLM Ops (MLOps for LLMs):** Covers the infrastructure and production-grade serving solutions necessary for deploying and managing LLMs, including topics like quantization and efficient inference

- **Prompt Engineering and Alignment:** Dedicated guides on the art and science of designing effective prompts and fine-tuning models to be helpful and harmless

Important Foundational Papers

Paper Title	Year	Core Contribution	Relevant LLM Terminology
Attention Is All You Need	2017	Introduced the **Transformer Architecture**, the pivotal structure for all modern LLMs.	Transformer Architecture
Outrageously Large Neural Networks: The Sparsely-Gated Mixture-of-Experts Layer	2017	The foundational paper for **Mixture-of-Experts (MoE)**, enabling models to scale in capacity while maintaining low inference cost by activating only a subset of experts.	Mixture-of-Experts (MoE)

(*continued*)

Paper Title	Year	Core Contribution	Relevant LLM Terminology
BERT: Pre-training of Deep Bidirectional Transformers for Language Understanding	2018	Popularized the **Encoder-Only** model and masked language modeling for deep language understanding.	Encoder-Only
Language Models are Few-Shot Learners	2020	Introduced GPT-3 and demonstrated the power of **In-Context Learning**, showcasing that very large models can perform tasks with **Few-Shot Prompting**.	Few-Shot Prompting, In-Context Learning
Retrieval-Augmented Generation for Knowledge-Intensive NLP Tasks	2020	Introduced **RAG (Retrieval-Augmented Generation)**, an architecture that grounds an LLM's answer in external, verifiable data to improve accuracy and freshness.	RAG (Retrieval-Augmented Generation)
LoRA: Low-Rank Adaptation of Large Language Models	2021	Introduced **LoRA**, a highly effective Parameter-Efficient Fine-Tuning (PEFT) method that significantly reduces computational cost and memory footprint during specialization.	LoRA, PEFT
Chain-of-Thought Prompting Elicits Reasoning in Large Language Models	2022	Demonstrated that including intermediate reasoning steps (the "chain of thought") in the prompt drastically improves the LLM's ability to solve complex, multi-step reasoning problems.	Chain-of-Thought (CoT)
Training Language Models to Follow Instructions with Human Feedback (InstructGPT)	2022	Detailed the process of **Instruction Fine-Tuning (IFT)** and using Reinforcement Learning from Human Feedback (RLHF) to achieve **Alignment** with human preferences.	Alignment, Instruction Fine-Tuning

(continued)

Paper Title	Year	Core Contribution	Relevant LLM Terminology
ReAct: Synergizing Reasoning and Acting in Language Models	2022	Proposed the **ReAct Framework**, formalizing the loop of Thought, Action (tool use), and Observation for building sophisticated **agentic workflows**.	ReAct Framework, Agentic Workflows

Index

CPU, *see* Central processing unit (CPU)

Cross-encoder model, 327

Cross-modal alignment, 343

Customer service, 6

H

I, J

K

L

M

V

Vector database, 326
Vector store, 321, 326
Vision-language models (VLMs), 377
    architectures
        BLIP, 344
        CLIP, 343
        LLaVA, 344, 345
    components, 341, 342
    implementation procedures
        automated image captioning with
          BLIP, 345–347
        VQA with LLaVA, 348–351
        zero-shot image classification,
          CLIP, 351–355
    projector, 342
Vision transformer (ViT), 341
Visual embeddings, 341, 342
Visual question answering (VQA),
        344, 345
    LLaVA, 348–351
Visual tokens, 342
ViT, *see* Vision transformer (ViT)
vLLM
    Docker container, 285–290
    FastAPI, 274–278
    Gradio objective, 281–285
    high-throughput serving
        workflow, 278–280
    memory fragmentation, 272
    OpenAI API, 273
    vs. TGI, 273
VLMs, *see* Vision-language
        models (VLMs)
VQA, *see* Visual question answering (VQA)

W, X, Y

Web scraping, 44
Weights & biases (W&B), 8
WordPiece, 49

Z

Zero-shot classification, 35, 343, 351–355
Zero-shot CoT, 81
Zero-shot/few-shot learning, 5
Zero-shot inference paradigm, 357
Zero-shot prompting, 69–75